The Sassafras Guide to Astronomy

Written by Paige Hudson

THIS PRODUCT IS INTENDED FOR HOME USE ONLY

The images and all other content in this book are copyrighted material owned by Elemental Science, Inc. Please do not reproduce this content on e-mail lists or websites. If you have an eBook, you may print out as many copies as you need for use WITHIN YOUR IMMEDIATE FAMILY ONLY. Duplicating this book or printing the eBook so that the book can then be reused or resold is a violation of copyright.

Schools and co-ops: You MAY NOT DUPLICATE OR PRINT any portion of this book for use in the classroom. Please contact us for licensing options at support@elementalscience.com.

THE SASSAFRAS GUIDE TO ASTRONOMY

First Edition 2020
Copyright @ Elemental Science, Inc.
Email: support@elementalscience.com

ISBN # 978-1-935614-94-4

Printed In USA For World Wide Distribution

For more copies write to :
Elemental Science
PO Box 79
Niceville, FL 32588
support@elementalscience.com

COPYRIGHT POLICY

All contents copyright ©2020 by Elemental Science. All rights reserved.

Limit of Liability and Disclaimer of Warranty: The publisher has used its best efforts in preparing this book, and the information provided herein is provided "as is." Elemental Science makes no representation or warranties with respect to the accuracy or completeness of the contents of this book and specifically disclaims any implied warranties of merchantability or fitness for any particular purpose and shall in no event be liable for any loss of profit or any other commercial damage, including but not limited to special, incidental, consequential, or other damages.

Trademarks: This book identifies product names and services known to be trademarks, registered trademarks, or service marks of their respective holders. They are used throughout this book in an editorial fashion only. In addition, terms suspected of being trademarks, registered trademarks, or service marks have been appropriately capitalized, although Elemental Science cannot attest to the accuracy of this information. Use of a term in this book should not be regarded as affecting the validity of any trademark, registered trademark, or service mark. Elemental Science is not associated with any product or vendor mentioned in this book.

The Sassafras Guide to Astronomy
Table of Contents

INTRODUCTION 5

    Book List   7
    Demonstration Supplies Listed By Chapter   11
    Project and Activity Supplies Listed By Chapter   13
    Guide to the Characters   14

CHAPTER LESSONS 17

    Chapter 1: Schedules   18
    Chapter 1: Time to Boldly go where...   20

    Chapter 2: Schedules   24
    Chapter 2: To Alaska and Beyond...   26

    Chapter 3: Schedules   30
    Chapter 3: Zipping Out of this World   32

    Chapter 4: Schedules   36
    Chapter 4: Space Sandwiches and Planetary Potato Chips   38

    Chapter 5: Schedules   42
    Chapter 5: The Rest of the Inner Planetary Scoop   44

    Chapter 6: Schedules   48
    Chapter 6: The Red Planet Rescue Mission   50

    Chapter 7: Schedules   54
    Chapter 7: Where in the solar system is Mr. Frye?   56

    Chapter 8: Schedules   60
    Chapter 8: Back To Earth   62

    Chapter 9: Schedules   66
    Chapter 9: Hawaiian Dreams and Sightings   68

    Chapter 10: Schedules   72
    Chapter 10: The National Air and Space Museum   74

    Chapter 11: Schedules   78
    Chapter 11: A Ballistic Heist   80

    Chapter 12: Schedules   84
    Chapter 12: LINLOC Failure to Launch   86

    Chapter 13: Schedules   90
    Chapter 13: The Copernicus Code   92

    Chapter 14: Schedules   96
    Chapter 14: The New Zealand Space Games   98

Chapter 15: Schedules   102
Chapter 15: The Planet Prowess Leaderboard   104
Chapter 16: Schedules   108
Chapter 16: The Set of Star Check   110
Chapter 17: Schedules   114
Chapter 17: The Orion Twins in Bollywood   116
Chapter 18: Schedules   120
Chapter 18: Back to Uncle Cecil's   122

Appendix 125

Lab Report Sheet   127
Paper Mâché Planet Model Directions   129
Venus/Earth Venn Diagram   130
Phases of the Moon   131
Scientist Biography Questionnaire   132
Constellation Templates   134

Glossary 135

Quizzes 139

Astronomy Quiz #1   141
Astronomy Quiz #2   143
Astronomy Quiz #3   145
Astronomy Quiz #4   147
Astronomy Quiz #5   149
Astronomy Quiz #6   151
Astronomy Quiz #7   153
Astronomy Quiz #8   155

THE SASSAFRAS GUIDE TO ASTRONOMY
INTRODUCTION

Our Living Books method of science instruction was first proposed in *Success in Science: A Manual for Excellence in Science Education*. This approach is centered on "living books" that are augmented by notebooking and scientific demonstrations. The students read (or are read to) from a science-oriented living book, such as *The Sassafras Science Adventures Volume 6: Astronomy*. Then they write about what they have learned and complete a related scientific demonstration or hands-on project. If time and interest allow, the teacher can add in non-fiction books that coordinate with the topic, or the students can do an additional activity and memorize related information.

The books of the *Sassafras Science Adventures* series are designed to give you the tools you need to employ the Living Books method of science instruction with your elementary students. For this reason, we have written an activity guide and logbook to correspond with each novel. This particular activity guide contains eighteen chapters of activities, reading assignments, scientific demonstrations, and so much more for studying astronomy.

Each of the chapters in this guide corresponds directly with the chapters in *The Sassafras Science Adventures Volume 6: Astronomy*. They are meant to give you the information you need to turn the adventure novel into a full science course for your elementary students. The chapters will provide you with a buffet of options that you can use to teach your students about the planets, the stars, and more. So pick and choose what you know you and your students will enjoy!

WHAT EACH CHAPTER CONTAINS

Each chapter begins with two schedule sheets for the corresponding chapter in *The Sassafras Science Adventures Volume 6: Astronomy*. On the schedule sheets, you will find a chapter summary, plus an overview of the supplies you will need for the demonstration, projects, and activities for the chapter. After that, you will find the optional schedules – one laid out as a four-day grid schedule and one laid out as a list to check off. These schedules are included to give you an idea of how your week could be organized, so please feel free to alter them to suit your needs.

After the scheduling information, you will find the information for the reading, notebooking, and activities for the particular chapter. This information is divided into the following sections:

SCIENCE-ORIENTED BOOKS
- **CHAPTER ASSIGNMENT** – This section contains the corresponding chapter in *The Sassafras Science Adventures Volume 6: Astronomy*.

- **ENCYCLOPEDIA READINGS** – This section contains possible reading assignments from:
    - *Basher Science Astronomy* (best for 1st through 2nd grades)
    - *Usborne Children's Encyclopedia* (best for 2nd through 4th grades)
    - *DK First Space Encyclopedia* (best for 2nd through 4th grades)
    - *Kingfisher Science Encyclopedia* (best for 4th through 6th grades)

  You can choose to read the assignments to the students or have the students read them on their own.

- **ADDITIONAL LIVING BOOKS** – This section contains a list of books that coordinate with what is being studied in the chapter. You can check these books out of your local library.

NOTEBOOKING
- **SCIDAT LOGBOOK INFORMATION** – This section has the information that the students could include in their SCIDAT logbook. It contains possible astronomical information the students could include on their astronomy record sheets. The students may or may not have all the same information

on their notebooking sheets, which is fine. You want their SCIDAT logbook to be a record of what they have learned. The information included is meant for you to use as a guide as you check their work. For more information about notebooking, please read the following articles:
- What is notebooking? – https://elementalscience.com/blogs/news/what-is-notebooking
- How to use notebooking with different ages – https://elementalscience.com/blogs/news/notebooking-with-different-ages

◯ VOCABULARY – This section includes vocabulary words that coordinate with each chapter. If your students are older, I recommend that you have them create a glossary of terms using a blank sheet of lined paper or the glossary sheets provided in *The Official Sassafras Student SCIDAT Logbook: Astronomy Edition*. You can also have them memorize these words and their definitions.

Scientific Demonstrations or Observations

☑ SCIENTIFIC DEMONSTRATION – This section includes a list of materials, the instructions, and an explanation for a scientific demonstration that coordinates with the chapter. A blank lab report sheet is provided for you in the Appendix on pp. 132-133 if you wish your students to write up the demonstration. If your students are in fourth grade or higher, I recommend that they complete at least one of these lab reports for this course.

Multi-week Projects or Activities

✂ ADDITIONAL ACTIVITIES – This section contains additional activities that go along with the chapter. There are multi-week projects, which will be done over several chapters, and activities that coordinate with that specific chapter. Pick and choose the activities that interest you and your students.

Memorization

☞ COPYWORK AND DICTATION – This section contains a short copywork passage and a longer dictation passage for you to use. Some students may use the shorter passages for dictation or the longer passages for copywork. Feel free to tailor the selections to your students' abilities. You can also use the selections as memory work assignments for the students.

Additional Materials

The back of this guide contains a few additional materials for your convenience. The first is a glossary of terms, which you can use with your students as they define the words for each chapter. After that, you will find a set of eight simple quizzes that you can use with your students to verify if your students are retaining the material.

A Word About the SCIDAT Logbook

The SCIDAT logbook is meant to be a record of your students' journey through their study of astronomy. It is explained in more detail in Chapter 1 of this guide. You can choose to make your own or purchase a pre-made logbook from Elemental Science. *The Official Sassafras SCIDAT Logbook: Astronomy Edition* has all the pages the students will need to create their own logbooks. Each one has been attractively illustrated for you so that you don't have to track down pictures for the students to use. This way, the students are able to focus on the information they are learning.

Final Thoughts

As the author and publisher of this curriculum, I encourage you to contact me at support@elementalscience.com with any questions or problems that you might have concerning *The Sassafras Guide to Astronomy*. I will be more than happy to answer them as soon as I am able. I hope that you and your students enjoy your journey through the world of astronomy with the Sassafras twins.

Book List

Main Text

The following book is required reading for the activities suggested in this guide.

- *The Sassafras Science Adventures Volume 6: Astronomy*

Encyclopedia Readings

The following encyclopedias have suggested pages scheduled in this guide. I recommend that you choose the one that best suits the age and ability of your students.

- *Basher Science Astronomy* (best for 1st through 2nd grades)
- *Usborne Children's Encyclopedia* (best for 2nd through 4th grades)
- *DK First Space Encyclopedia* (best for 2nd through 4th grades)
- *Kingfisher Science Encyclopedia* (best for grades 4th through 6th grades)

You may want to consider purchasing the following resource for your night sky spotting to use for the remainder of your astronomy studies:

- *100 Things to Spot in the Night Sky (Spotter's guides) Cards*

Recommended Resources

The following book will be very beneficial to have when completing this course. It contains all the pages and pictures your students will need to record their journey through astronomy.

- *The Official Sassafras Student SCIDAT Logbook: Astronomy Edition*

View all the links mentioned in this guide in one place and get a digital copy of the templates, glossary, and quizzes by visiting the following page:

- http://sassafrasscience.com/volume-6-links/

Additional Living Books Listed By Chapter

Chapter 1
- *A Cat's Guide to the Night Sky* by Stuart Atkinson and Brendan Kearney
- *Space: A Visual Encyclopedia* by DK

Chapter 2
- *There's No Place Like Space: All About Our Solar System (Cat in the Hat's Learning Library)* by Tish Rabe and Aristides Ruiz
- *Scholastic Reader Level 2: Solar System* by Gregory Vogt
- *Magic School Bus Out of This World : A Book about Space Rocks* by Joanna Cole and Bruce Degen

Chapter 3
- *The Sky Is Full of Stars (Let's-Read-and-Find-Out Science 2)* by Franklyn M. Branley and Felicia Bond
- *Jump Into Science: Stars* by Steve Tomecek
- *Stars! Stars! Stars!* by Bob Barner
- *If You Were a Kid Docking at the International Space Station (If You Were a Kid)* by Josh Gregory and Jason Raish

- *International Space Station (Let's-Read-and-Find-Out Science 2)* by Dr. Franklyn M. Branley and True Kelley

CHAPTER 4
- *Mercury (First Facts: Solar System)* by Adele Richardson
- *Mercury (Scholastic News Nonfiction Readers: Space Science)* by Christine Taylor-Butler
- *Mercury and Venus (Up in Space)* by Rosalind Mist
- *Venus (True Books: Space)* by Elaine Landau
- *Venus (First Facts: Solar System)* by Adele Richardson
- *Venus (Scholastic News Nonfiction Readers: Space Science)* by Melanie Chrismer

CHAPTER 5
- *Planet Earth/Inside Out* by Gail Gibbons
- *The Magic School Bus Inside the Earth (Magic School Bus)* by Joanna Cole and Bruce Degen
- *See Inside Planet Earth (Usborne Flap Book)* by Katie Daynes and Peter Allen
- *You Are the First Kid on Mars* by Patrick O'Brien
- *Mission to Mars (Let's-Read-and-Find-Out Science 2)* by Franklyn M. Branley and True Kelley
- *Mars (Scholastic News Nonfiction Readers: Space Science)* by Melanie Chrismer

CHAPTER 6
- *Destination: Jupiter* by Seymour Simon
- *Jupiter: The Largest Planet (Our Solar System)* by Daisy Allyn
- *Planet Jupiter (True Books)* by Ann O. Squire
- *Jupiter (Scholastic News Nonfiction Readers: Space Science)* by Christine Taylor-Butler
- *Jupiter (Blastoff! Readers: Exploring Space)* by Derek Zobel
- *Saturn* by Seymour Simon
- *Saturn (True Books)* by Elaine Landau
- *Saturn: The Ringed Planet (Our Solar System)* by Daisy Allyn
- *Planet Saturn (True Books)* by Ann O. Squire
- *Saturn (Scholastic News Nonfiction Readers: Space Science)* by Christine Taylor-Butler
- *Jupiter and Saturn (Up in Space)* by Rosalind Mist

CHAPTER 7
- *Uranus (True Books)* by Christine Taylor-Butler
- *Uranus* by Seymour Simon
- *Uranus: The Ice Planet (Our Solar System)* by Greg Roza
- *The Sideways Planet: Uranus (Amazing Science: Planets)* by Nancy Loewen and Jeff Yesh
- *Neptune* by Seymour Simon
- *Neptune: The Stormy Planet (Our Solar System)* by Greg Roza
- *Planet Neptune (True Books)* by Ann O. Squire
- *Neptune (Scholastic News Nonfiction Readers: Space Science)* by Melanie Chrismer
- *Farthest from the Sun: The Planet Neptune (Amazing Science: Planets)* by Nancy Loewen and Jeff Yesh

CHAPTER 8
- *The Milky Way (Exploring Space)* by Martha E. H. Rustad and Ilia I. Roussev
- *The Milky Way (Galaxy)* by Gregory L. Vogt
- *Looking Through a Telescope (Rookie Read-About Science)* by Linda Bullock
- *Telescopes (First Facts: Science Tools)* by Adele Richardson
- *The Hubble Telescope (Blastoff! Readers: Exploring Space)* by Derek Zobel

Chapter 9
- *Satellites and Space Probes (Eye on the Universe)* by Niki Walker
- *All About Satellites (Blast Off!)* by Miriam Gross
- *Satellites (Let's See Library: Communication)* by Darlene R. Stille

Chapter 10
- *The Sun: Our Nearest Star (Let's-Read-and-Find-Out)* by Franklyn M. Branley and Edward Miller
- *The Sun* by Seymour Simon
- *Day and Night (First Step Nonfiction: Discovering Nature's Cycles)* by Robin Nelson
- *Day and Night (Patterns in Nature)* by Margaret Hall and Jo Miller
- *What Makes Day and Night (Let's-Read-and-Find-Out Science 2)* by Franklyn M. Branley and Arthur Dorros

Chapter 11
- *Faces of the Moon* by Bob Crelin and Leslie Evans
- *The Moon Book* by Gail Gibbons
- *The Moon Seems to Change (Let's-Read-and-Find Out Science 2)* by Franklyn M. Branley and Barbara and Ed Emberley
- *Eclipses (Amazing Sights of the Sky)* by Martha Elizabeth Hillman Rustad
- *Eclipse Chaser: Science in the Moon's Shadow (Scientists in the Field Series)* by Ilima Loomis and Amanda Cowan

Chapter 12
- *Mousetronaut: Based on a (Partially) True Story (Paula Wiseman Books)* by Mark Kelly and C. F. Payne
- *If You Decide to Go to the Moon* by Faith McNulty and Steven Kellogg
- *DK Readers L2: Astronaut: Living in Space* by Deborah Lock
- *Floating in Space (Let's-Read-and-Find-Out Science 2)* by Franklyn M. Branley and True Kelley
- *DK Readers L1: Rockets and Spaceships* by Karen Wallace

Chapter 13
- *Comets, Meteors, and Asteroids* by Seymour Simon
- *Asteroids, Comets, and Meteorites (First Facts)* by Steve Kortenkamp
- *Asteroids and Comets (Science Readers: A Closer Look)* by William B. Rice
- *Comets and Asteroids: Space Rocks (Our Solar System)* by Greg Roza

Chapter 14
- *Spacewalks (Little Astronauts)* by Kathryn Clay
- *Space Walks (Our Solar System)* by Dana Meachen Rau and Nadia Higgins
- *Endurance, Young Readers Edition: My Year in Space and How I Got There Paperback* by Scott Kelly
- *DK Readers L2: Spaceships and Rockets: Relive Missions to Space (DK Readers Level 2)* by DK
- *Go for the Moon: A Rocket, a Boy, and the First Moon Landing* by Chris Gall

Chapter 15
- *When Is a Planet Not a Planet?: The Story of Pluto* by Elaine Scott
- *Pluto: Dwarf Planet (Scholastic News Nonfiction Readers: Space Science)* by Christine Taylor-Butler
- *Dwarf Planets: Pluto, Charon, Ceres, and Eris (Amazing Science: Planets)* by Nancy Loewen and Jeff Yesh
- *Black Holes (A True Book: Space)* by Ker Than
- *A Black Hole Is Not a Hole* by Carolyn Cinami DeCristofano
- *What's Inside a Black Hole? Theories About Space Phenomena (Beyond the Theory: Science of the Future)* by Tom Jackson

Chapter 16 and 17
- 📖 *Glow in the Dark Constellations* by C. E. Thompson
- 📖 *What We See in the Stars: An Illustrated Tour of the Night Sky* by Kelsey Oseid
- 📖 *Zoo in the Sky: A Book of Animal Constellations* by Jacqueline Mitton
- 📖 *Once Upon a Starry Night: A Book of Constellations* by Jacqueline Mitton and Christina Balit

Chapter 18
- 📖 *The Three Little Aliens and the Big Bad Robot* by Margaret McNamara and Mark Fearing
- 📖 *Aliens in Underpants Save the World (The Underpants Books)* by Claire Freedman and Ben Cort
- 📖 *Your Alien* by Tammi Sauer and Goro Fujita

Demonstration Supplies Listed By Chapter

Chapter 1: Observing the Night Sky
No supplies needed

Chapter 2: Impact
Marbles
Aluminum pan
Cornstarch
Cocoa Powder
Tape measure

Chapter 3: Shining Stars
Black construction paper
Toothpick
Tape
Flashlight
Large clear bowl
Water

Chapter 4: Trapped Heat
2 Thermometers
Cutting board
Clear glass bowl

Chapter 5: Solar Rover
A solar-powered mini-car kit
(OR a DC motor, Solar panel with wires, 2 Sets of wheels with axles or steel wires, Cardboard, Hot glue, Plastic tubing)

Chapter 6: Stormy Swirls
Bowl
Milk
Food Coloring
Liquid soap
Toothpick

Chapter 7: Planetary Orbit
Marble
Smooth pie plate or cake pan

Chapter 8: Magnify
Glass bowl
Cooking oil
Piece of paper with words on it
Magnifying glass

Chapter 9: Reflection Direction
Small mirror
Small flashlight
A dark room

Chapter 10: Solar S'mores
Large marshmallows
Chocolate squares
Graham crackers
Foil
Cardboard box
Plastic wrap

Chapter 11: Moon Cookies
8 Sandwich-style cookies
Picture of the phases of the moon (*Template is found on Appendix pg. 131.*)

Chapter 12: Space Tasks
Thick yellow rubber gloves or work gloves
LEGO bricks
Several bolts, washers, and nuts

Chapter 13: Simple Astrolabe
Thin wooden dowel or a straw
String (about 12" long)
Heavy metal nut or washer
Protractor
Tape

Chapter 14: Balloon Rocket
Straw
String (5 feet)
Scissors
Large balloon
2 Chairs
Tape

Chapter 15: Sucked In
Hard-boiled egg
Warm water
Bottle with large-mouth (i.e., sports drink bottle)
Access to a freezer

Chapter 16 And 17: Flashlight Planetarium
Foil
Toilet Paper Tube
Pin
Small flashlight
Constellation pictures (*Templates are found on Appendix pg. 134.*)
Rubber band
Sharpie marker

Chapter 18:
Planetary Bingo Cards (*Download these for free from Elemental Science.*)

Project and Activity Supplies Listed By Chapter

The projects and activities listed in this guide are optional, so you may not need all of these supplies. However, this list has been provided for your convenience. If you do decide to do these projects, in addition to the items listed each week you will need clear tape, glue, scissors, a variety of paint colors, and a set of markers.

Chapter 1
Black construction paper (2-11"x 17" or 3-8 ½" x 11" sheets)
Clear gel glue, Water, Silver glitter, Black (or purple) food coloring, Plastic baggie, Cup, Borax Laundry Booster

Chapter 2
Brown and yellow construction paper
Beads, String
Ice cream, Plastic baggie, Rolling pin, Cookies
Rocks, Black paint, Silver glitter

Chapter 3
Black construction paper, 2 Cotton balls, 1 Small yellow pom-pom, 1 Large red pom-pom, 1 Large white sequin

Chapter 4 to 7
Materials will vary based on the type of planet model you choose to make.

Chapter 8
White tissue paper or white chalk pastel
White chalk pastel or crayon, black construction paper, silver glitter

Chapter 9
Bottle caps, Toothpicks, Thin cardboard, A small juice box, Glittered blue decorative card stock, Gold and silver paint, 12x12-inch Foam piece, Aluminum foil, Glue, 1/4-inch Wooden dowel, Scissors, Pencil, Measuring tape (Or LEGO bricks)
Smartphone

Chapter 10
White glue, Food coloring, Toothpicks, Yogurt container lid, Hole punch, String
Colored pencils or magazine pictures

Chapter 11
Black poster board, White toothpaste or shaving cream, Butter knife, Tape, Wiffle ball
2 Sheets of paper, Scissors, Yellow and orange paint, Paintbrush

Chapter 12
Soda bottle, Cardstock, White and black paint, Glue
String, Small paper cup, Mini-marshmallows, Potential parachute material (paper, tissue, thin fabric, felt), Scissors or a hole punch

Chapter 13
3' Curling ribbon, Tennis ball, Foil, Straight pin

Chapter 14
Build-a-rocket kit

Chapter 15
Materials will vary based on the type of planet model you choose to make.
Plastic cup, Sharpie markers, Pan, Spray oil, Foil

Chapter 16
Marshmallows, Toothpicks
Gold star stickers, White crayon, Paper, Dark blue or black paint

Chapter 17
Materials will vary based on what you decided to do for your constellation party.

Chapter 18
Materials will vary based on what you decided to do for the alien craft.

The Sassafras Guide to the Characters Found in Volume 6: Astronomy

Old Friends that Appear Throughout the Book*

- **Blaine Sassafras** – A.K.A. Train and Rowboat, this boy is the male twin of the soon-to-be-famous Sassafras twins. He also has a wide range of acting abilities and a talent for finding things in space.
- **Tracey Sassafras** – A.K.A. Blaisey and Fish Hook, this girl is the female twin of the soon-to-be-famous Sassafras twins. She also plays a mean game of Pass the Petri Dish and Copernicus Code.
- **Uncle Cecil** – The Sassafras twins' talented, eccentric, and messy uncle who can never get their names right. He is an inventor with friends and colleagues that are out of this world.
- **President Lincoln** – A.K.A. Linc Dog and The Prez, this prairie dog is Uncle Cecil's lab assistant. He doesn't say much, but without his talent for inventing, the twins' adventures wouldn't be the same.
- **Summer Beach** – The sandwich-loving, excitable scientist who is a dear friend of Uncle Cecils. Her lab in Alaska is packed with out of this world tech!
- **Ulysses S. Grant** – Summer's lab assistant, who happens to be an Arctic ground squirrel, and best friend of President Lincoln. He invents all kinds of technology, such as robot squirrels and zip-zop cuffs, when he isn't hibernating.
- **The Man With No Eyebrows** – The memory-erasing, disappearing cape-wearing, eyebrow-less man who has tried just about everything he can think of to stop the twins. In the last volume we found out his name: Thaddeus.
- **Adrianna Archer** – The former Triple S agent who the twins met during their Earth Science leg. She worked with the Man With No Eyebrows behind the scenes during their Geology leg.
- **Evan DeBlose** – A.K.A. Agent Beans, this Triple S agent who serves as the twin's local expert on their Earth Science. He was friends and colleagues with Adrianna Archer until she went rogue.
- **Q-Tip** – A Triple S's agent and resident expert in technologizing. The twins got a glimpse of his talents on their Earth Science leg.
- **Captain Marolf** – Head of the Triple S Agency. The twins met the captain during the Earth Science leg.
- **Yuroslav Bogdanovich** – The rogue scientist who bears an uncanny resemblance to Uncle Cecil. He tried to stop the twins during their Earth Science leg, but his memory was erased during the twin's Geology leg and now he works as a clerk at the Left-Handed Turtle Market.

Cecil's Neighborhood (Chapter 1)

- **The Guardian Beast (in name only)** – This miniature poodle is the stuff nightmares are made of, according to Cecil Sassafras.
- **Old Man Grusher** – Uncle Cecil's neighbor and owner of the Guardian Beast.

Chapters 2 & 3 (Summer's Alaskan Lab)

- **REESE** – The robotic creation of President Lincoln and Ulysses S. Grant. His name stands for Robotic Exploration, Entertainment, and Scientific Enhancement. REESE joins the twins at several other locations throughout the book.

Chapters 4 to 7 (International Space Station)

- **Yang Bo** – The twins' local expert in space. He is a Chinese astronaut living on the International Space Station, serving as the station's astrobiologist. He is also a former classmate of Uncle Cecil and Summer.
- **Captain Dianna Sturgess** – The decorated American astronaut who is currently in charge of the International Space Station.
- **Sander Petrov** – The Russian astronaut who serves as the International Space Station mechanic.

- ⭐ **Bayard Clemence** – The French astronaut who serves as the International Space Station physician.
- ⭐ **Anna Maria Bezerra** – The Brazilian astronaut who serves as the I.S.S. meteorologist.
- ⭐ **Parth Banerjee** – The Indian astronaut who serves as the International Space Station mathematician.
- ⭐ **SLIM** – The QA-700 robot aboard the International Space Station Its name stands for Super Literal Information Machine.
- ⭐ **Brett Frye** – The billionaire space tourist who spent some time on the International Space Station.
- ⭐ **Queenie Clemence (in name only)** – The sister of Bayard Clemence.

Chapters 8 & 9 (Hawaii)

- ⭐ **J.P. Jungos** – The twins' local expert for their time in Hawaii. He is a burn victim on the journey of his lifetime.
- ⭐ **Peter Karko** – An employee at the Mauna Kea Observatories, who also turns out to be a very opportunistic and cranky tour guide.
- ⭐ **Dr. Ellison Ocampo** – Lead scientist at the Mauna Kea Observatories.

Chapters 10 & 11 (Washington, D.C.)

- ⭐ **Paul Sims** – The twins' local expert as they explore the National Air and Space Museum. He is a very knowledgeable museum curator, but there is more to him than meets the eye.
- ⭐ **Wiggles and Fidget** – The excitable museum security guards.
- ⭐ **Sparks Sheen** – The leader of the family custodial crew, Shine-O-Mite.
- ⭐ **Flash Sheen** – A member of the Shine-O-Mite crew.
- ⭐ **Pat Sheen** – Short for Patina, she is the only sister on the Sheen Shine-O-Mite crew.
- ⭐ **Lumin Sheen** – A member of the Shine-O-Mite crew.
- ⭐ **Alexander Slote** – A member of the Rotary Club – not the one you are thinking, it's a club extoling the virtues of the rotary telephone.
- ⭐ **Graham Slote** – Another member of the Rotary Club.
- ⭐ **Belle Slote** – The final member of the Rotary Club. She is the only female in the three-member group.

Chapters 12 & 13 (Poland)

- ⭐ **Minka Ziven** – The twins' local expert for their time in Poland and the clue guide for the Copernicus Code Escape Room.
- ⭐ **Clive Stanek** – One of the players at the Copernicus Code Escape Room. He is Halley's brother.
- ⭐ **Halley Stanek** – Another one of the players at the Copernicus Code Escape Room. She is Clive's sister.

Chapters 14 & 15 (New Zealand)

- ⭐ **Arty Stone** – He is the lead foreman for the Professional Gamer Championships and the twins' local

- ⭐ **Mr. Sebastian** – The production manager and announcer for the Professional Gamer Championships.
- ⭐ **Wayne Hammer** – The boisterous American gamer competing in the Professional Gamer Championships.
- ⭐ **Ms. Pink Rocker** – The pink-clad Russian gamer competing in the Professional Gamer Championships.
- ⭐ **Mohawk Wellington** – The preppy, Mohawk-wearing British gamer competing in the Professional Gamer Championships.
- ⭐ **Agnes the Librarian** – The quiet, middle-aged Canadian gamer competing in the Professional Gamer Championships.
- ⭐ **El Cohete Loco (The Crazy Rocket)** – The easy-going Mexican gamer competing in the Professional Gamer Championships.
- ⭐ **Robbie Thistler** – The mysterious gamer who set an unbeatable record on Planet Prowess before disappearing from the gaming scene.
- ⭐ **Kiwi Jones** – The young gamer who ends up shows up all of the top five gamers at the Professional Gamer Championships.

expert for their time in New Zealand.

CHAPTERS 16 & 17 (MUMBAI, INDIA)

- ⭐ **Ravi Chopraz** – The twins' local expert for their time in India and star of Bollywood's famous *Star Check* show. He plays the handsome Captain Cutta.
- ⭐ **Varun Gowda** – The director of *Star Check*.
- ⭐ **Preathi** – The digital effects specialist for *Star Check*.
- ⭐ **Sana and Puji** – Two of the stylists for *Star Check*.
- ⭐ **Jaya Amin** – She plays First Lieutenant Ursa, the captain's wiser and better-looking counterpart.
- ⭐ **Rom Basu** – He plays the factual and stoic Second Lieutenant Denab.
- ⭐ **Chiku Kapadia** – He plays a Worbflyster from the planet Worbflyse, who only speaks Worbflystian.
- ⭐ **Dhruv Dalal** – The resident *Star Check* stunt man, who also plays the part of a Borbothian alien who is out to get the captain.
- ⭐ **Aja and Ru Katri** – Executives from the company that is responsible for producing *Star Check*.

CHAPTER 18 (BACK AT UNCLE CECIL'S LAB)

- ⭐ **Socrates and Aristotle** – The skeletons-turned-mannequins from the twins' Anatomy leg that Cecil frequently talks to as if they are friends.

CHAPTER LESSONS

Chapter 1: Grid Schedule

Supplies Needed	
Demo	• No supplies needed
Projects	• Black construction paper (2-11"x 17" or 3-8 ½" x 11" sheets), Clear gel glue, Water, Silver glitter, Black (or purple) food coloring, Plastic baggie, Cup, Borax laundry booster

Chapter Summary

The chapter opens with a dilemma: Uncle Cecil must risk going into Old Man Grusher's backyard, possibly facing off with his Guardian Beast, in order to retrieve the petri dish, otherwise known as a frisbee, that landed there during a game of Pass the Petri. After several pep talks, he finally makes it over the fence with the help of Blaine and Tracey. They meet Old Man Grusher and learn that he is not as bad as they thought before heading back to Cecil's basement lab where the twins get to hear President Lincoln's ever-so-brief presentation on geology. We learn that the next leg of the twin's journey is Astronomy, and Summer is going to be their local expert. Before the chapter closes, we also learn that the Man With No Eyebrows has not given up; in fact, he has a whole army of scientists helping him now, thanks to Adrienne Archer, the rough Swiss Secret Service agent!

Weekly Schedule

	Day 1	Day 2	Day 3	Day 4
Read	☐ Read the section entitled "Pass the Petri Gone Wrong" of Chapter 1 in *SSA* Volume 6: Astronomy*.	☐ (*Optional*) Read one or all of the assigned pages from the encyclopedia of your choice.	☐ Read the section entitled "A Look Back at Geology" of Chapter 1 in *SSA Volume 6: Astronomy*.	☐ (*Optional*) Read one of the additional books from your library.
Write	☐ Set up your students' SCIDAT logbooks.	☐ (*Optional*) Write a narration on the Astronomy Notes Sheet on SL** pg. 5. ☐ Add information learned from the demonstration on SL pg. 5.	☐ Go over the vocabulary word and enter it into the Astronomy Glossary on SL pg. 91.	☐ (*Optional*) Complete the copywork or dictation assignment and add it to the Astronomy Notes sheet on SL pg. 6.
Do	☐ (*Optional*) Play a game of "I Spy."	☐ Do the demonstration entitled "Observing the Night Sky."	☐ (*Optional*) Make Night Sky Slime.	☐ Work on the Solar System model.

*SSA = *The Sassafras Science Adventures*
**SL = *The Official Sassafras SCIDAT Logbook: Astronomy Edition*

Chapter 1: List schedule

	Supplies Needed
Demo	• No supplies needed
Projects	• Black construction paper (2-11"x 17" or 3-8 ½" x 11" sheets), Clear gel glue, Water, Silver glitter, Black (or purple) food coloring, Plastic baggie, Cup, Borax laundry booster

Chapter Summary

The chapter opens with a dilemma: Uncle Cecil must risk going into Old Man Grusher's backyard, possibly facing off with his Guardian Beast, in order to retrieve the petri dish, otherwise known as a frisbee, that landed there during a game of Pass the Petri. After several pep talks, he finally makes it over the fence with the help of Blaine and Tracey. They meet Old Man Grusher and learn that he is not as bad as they thought before heading back to Cecil's basement lab where the twins get to hear President Lincoln's ever-so-brief presentation on geology. We learn that the next leg of the twin's journey is Astronomy, and Summer is going to be their local expert. Before the chapter closes, we also learn that the Man With No Eyebrows has not given up; in fact, he has a whole army of scientists helping him now, thanks to Adrienne Archer, the rough Swiss Secret Service agent!

Essential To-Do's

Read
☐ Read the section entitled "Pass the Petri Gone Wrong" of Chapter 1 in *SSA* Volume 6: Astronomy*.
☐ Read the section entitled "A Look Back at Geology" of Chapter 1 in *SSA Volume 6: Astronomy*.

Write
☐ Set up your students' SCIDAT logbooks.
☐ Add information learned from the demonstration on SL** pg. 5.
☐ Go over the vocabulary word and enter it into the Astronomy Glossary on SL pg. 91.

Do
☐ Do the demonstration entitled "Observing the Night Sky."
☐ Work on the Solar System model.

Optional Extras

Read
☐ Read one or all of the assigned pages from the encyclopedia of your choice.
☐ Read one of the additional books from your library.

Write
☐ Write a narration on the Astronomy Notes Sheet on SL pg. 5.
☐ Complete the copywork or dictation assignment and add it to the Astronomy Notes sheet on SL pg. 6.

Do
☐ Play a game of "I Spy."
☐ Make Night Sky Slime.

*SSA = The Sassafras Science Adventures
**SL = The Official Sassafras SCIDAT Logbook: Astronomy Edition

Chapter 1: Time to Boldly Go Where...

Science-Oriented Books

Living Book Spine
- Chapter 1 of *The Sassafras Science Adventures Volume 6: Astronomy*

Optional Encyclopedia Readings
- *Basher Science Astronomy* pg. 4 (Introduction)
- *Usborne Children's Encyclopedia* pp. 246-247 (Amazing Space)
- *DK First Space Encyclopedia* pp. 4-5 (What is space?)
- *Kingfisher Science Encyclopedia* pg. 385 (Space and Time)

Additional Books
- *A Cat's Guide to the Night Sky* by Stuart Atkinson and Brendan Kearney
- *Space: A Visual Encyclopedia* by DK

Notebooking (SCIDAT Logbook Information)

This week, you will set up the students' SCIDAT logbooks. You can use blank sheets of copy paper with dividers for each section or purchase *The Official Sassafras Student SCIDAT Logbook: Astronomy Edition* with all the pages and pictures from Elemental Science. Below is an explanation of each of the student sheets.

Night Sky Journal Sheets

The purpose of these sheets is to give the students an opportunity to work on their spotting skills as they create a night sky journal throughout this leg of the journey.

- **Blank Space** – Have the students draw what they see or add a picture in the space above the boxes.
- **Date and Time** – Have the students add the date and time they made the observations they recorded on the night sky journal sheet.
- **Where We Were** – Have the students write down the location that they were at when they made the observations they recorded on the night sky journal sheet.
- **What We Saw** – Have the students enter the observations they have on the night sky journal sheet.

Astronomy Record Sheets

The purpose of these sheets is for the students to record what they have learned about the various topics that are introduced in *The Sassafras Science Adventures Volume 6: Astronomy*.

- **Information Learned** – The students should color the picture above the box, if they desire, and enter any information that they have learned about the particular topic.

Astronomy Science Notes Sheets

The purpose of these sheets is for the students to record any additional information that they have learned during their study of astronomy. You can use these sheets to record additional narrations, copywork, or dictation assignments.

Project Record Sheets

The purpose of these sheets is for the students to record the projects they have done during the course of their study of astronomy.

Astronomy Glossary

The purpose of the glossary is for the students to create a dictionary of terms that they have encountered while reading *The Sassafras Science Adventures Volume 6: Astronomy*. They can look up each term in a science encyclopedia or in the glossary included on pp. 136-137 of this guide. Then have the students copy each definition onto a blank index card or into their SCIDAT logbooks. They should also illustrate each of the vocabulary words. (**NOTE** – *In The Official Sassafras Student SCIDAT Logbook: Astronomy Edition, these pictures are already provided*.) This week, have the students look up the following terms:

- **ASTRONOMY** – The branch of science that studies what is out in space.

For each of these sheets, you can have the students enter information only from *The Sassafras Science Adventures Volume 6: Astronomy*, or you can have them do additional research to gather more facts. What you choose to do will depend on the ages and abilities of your students.

Scientific Demonstration: Observing The Night Sky

Begin by taking a moment to discuss things that you can see in the night sky, such as stars, planets, satellites, and the moon. You can also discuss how important observation skills are for the scientist who is studying astronomy. You can view the following blog posts and podcast for more information on the subject of observation:

- http://elementalscience.com/blogs/news/63858627-observation-is-key
- http://elementalscience.com/blogs/homeschool-science-tips/71117699-3-ways-to-work-on-observation
- https://elementalscience.com/blogs/podcast/episode-9

Explain that, today, the students are going to practice their observation skills while doing a bit of night sky spotting. Then, head outside and use a telescope or binoculars to look up at the night sky. Allow the students to make observations and ask questions. Ask the students:

? What do you see?

Allow the students to observe the night sky for a time. You can use apps like Google Sky (Android) or StarWalk (Apple) to help identify what you are seeing. Have the students look for constellations and planets, or just have them identify the phase of the moon. Record their observations on the sheet provided in the SCIDAT logbook or in a night sky journal, as explained below.

Multi-Week Projects And Activities

Multi-week Projects

- **SOLAR SYSTEM MODEL** – Over the weeks of this study, the students will create a large wall-sized solar system model or a smaller lap-sized construction-paper version. This week, you will need to get your model space ready. If you are going to do a wall version, pick out the wall you would like to use. If you are going to do the lap version, have the students tape together two 11"x 17" (or three 8 ½" x 11") sheets of black construction paper together to make an 11" x 34" (or 8 ½"x 33") sheet of paper.

Activities For This Week

- **I SPY** – Play a game of "I Spy" to help the students work on their observation skills.
- **NIGHT SKY SLIME** – Have the students make a batch of night sky slime! You will need clear gel glue, water, silver glitter, black (or purple) food coloring, a plastic baggie, a cup, and borax laundry booster. Begin by mixing 4 oz. of glue with 4 oz. of water, a few drops of food coloring, and a shake

of glitter in a plastic bag. Next, in a separate cup, mix a quarter cup of water with half a teaspoon of borax. Then, add the borax solution to the baggie and massage the bag for a few minutes until a nice firm slime has formed. Pull the slime out of the baggie and have fun!

Memorization

Copywork/Dictation

☞ **Copywork Selection**

Astronomers study what is out in space.

☞ **Dictation Passage**

Astronomy is the branch of science that studies what is out in space. Astronomers study planets, stars, black holes, galaxie,s and much more. They use telescopes, satellites, and space probes to learn about space.

Chapter 1 Notes

Chapter 2: Grid Schedule

	Supplies Needed
Demo	• Marbles, Aluminum pan, Cornstarch, Cocoa Powder, Tape measure
Projects	• Brown and yellow construction paper • Beads, String, Ice cream, Plastic baggie, Rolling pin, Cookies, Rocks, Black paint, Silver glitter

Chapter Summary

The chapter opens with Blaine, Tracey, Summer, and President Lincoln zipping to Summer's lab in Alaska. We then find out that the Man With No Eyebrows, dressed in the Dark Cape suit, is launching into space in his own personal craft (*Thad-1*) thanks to none other than Adrianna Archer. Then we switch to Agent DeBlose of the Swiss Secret Service, where we learn that he and Q-Tip are also being launched into space in the *Dauntless-12* to fix a satellite. Back at Summer's lab, the twins learn about the solar system as they re-tour her lab and realize that they have already been through the cockpit of her lab-spaceship. They meet REESE, the robot who will help them with their SCIDAT while they are in space. REESE shares information about asteroids and the chapter wraps up with a dance party, courtesy of the robot's song about gravity!

Weekly Schedule

	Day 1	**Day 2**	**Day 3**	**Day 4**
Read	☐ Read the section entitled "Rocketing in the Solar System" of Chapter 2 in *SSA Volume 6: Astronomy*.	☐ Read the section entitled "Summer's Spaceship and Asteroid-sharing Robots" of Chapter 2 in *SSA Volume 6: Astronomy*.	☐ (*Optional*) Read one or all of the assigned pages from the encyclopedia of your choice.	☐ (*Optional*) Read one of the additional books from your library.
Write	☐ Fill out a Astronomy Record Sheet on SL pg. 9 on the solar system. ☐ Go over the vocabulary words and enter them into the Astronomy Glossary on SL pg. 91-92.	☐ Fill out a Astronomy Record Sheet on SL pg. 10 on asteroids. ☐ (*Optional*) Add observations to the Night Sky Journal Sheet on SL pg. 7.	☐ (*Optional*) Write narration on the Astronomy Notes Sheet on SL pg. 13. ☐ Add information learned from the demonstration on SL pg. 13.	☐ (*Optional*) Complete the copywork or dictation assignment and add it to the Astronomy Notes sheet on SL pg. 13. ☐ (*Optional*) Fill out the record sheet on SL pg. 15 for one of the projects.
Do	☐ (*Optional*) Make the Solar System Bracelet.	☐ (*Optional*) Do the Asteroids and/or Meteors projects.	☐ Do the demonstration entitled "Impact."	☐ Work on the Solar System Model.

Chapter 2: List schedule

	Supplies Needed
Demo	• Marbles, Aluminum pan, Cornstarch, Cocoa Powder, Tape measure
Projects	• Brown and yellow construction paper • Beads, String, Ice cream, Plastic baggie, Rolling pin, Cookies, Rocks, Black paint, Silver glitter

Chapter Summary

The chapter opens with Blaine, Tracey, Summer, and President Lincoln zipping to Summer's lab in Alaska. We then find out that the Man With No Eyebrows, dressed in the Dark Cape suit, is launching into space in his own personal craft (*Thad-1*) thanks to none other than Adrianna Archer. Then we switch to Agent DeBlose of the Swiss Secret Service, where we learn that he and Q-Tip are also being launched into space in the *Dauntless-12* to fix a satellite. Back at Summer's lab, the twins learn about the solar system as they re-tour her lab and realize that they have already been through the cockpit of her lab-spaceship. They meet REESE, the robot who will help them with their SCIDAT while they are in space. REESE shares information about asteroids and the chapter wraps up with a dance party, courtesy of the robot's song about gravity!

Essential To-Do's

Read
- ☐ Read the section entitled "Rocketing in the Solar System" of Chapter 2 in *SSA Volume 6: Astronomy*.
- ☐ Read the section entitled "Summer's Spaceship and Asteroid-sharing Robots" of Chapter 2 in *SSA Volume 6: Astronomy*.

Write
- ☐ Fill out a Astronomy Record Sheet on SL pg. 9 on the solar system.
- ☐ Go over the vocabulary words and enter them into the Astronomy Glossary on SL pg. 91-92.
- ☐ Fill out a Astronomy Record Sheet on SL pg. 10 on asteroids.
- ☐ Add information learned from the demonstration on SL pg. 13.

Do
- ☐ Do the demonstration entitled "Impact."
- ☐ Work on the Solar System Model.

Optional Extras

Read
- ☐ Read one or all of the assigned pages from the encyclopedia of your choice.
- ☐ Read one of the additional books from your library.

Write
- ☐ Add observations to the Night Sky Journal Sheet on SL pg. 7.
- ☐ Write a narration on the Astronomy Notes Sheet on SL pg. 13.
- ☐ Complete the copywork or dictation assignment and add it to the Astronomy Notes sheet on SL pg. 13.
- ☐ Fill out the record sheet on SL pg. 15 for one of the projects.

Do
- ☐ Make the Solar System Bracelet.
- ☐ Do the Asteroids and/or Meteors projects.

Chapter 2: To Alaska and Beyond...

Science-Oriented Books

Living Book Spine
- Chapter 2 of *The Sassafras Science Adventures Volume 6: Astronomy*

Optional Encyclopedia Readings
- *Basher Science Astronomy* pg. 8 (Solar System), pg. 18 (Meteorite), pg. 28 (Asteroid Belt)
- *Usborne Children's Encyclopedia* pp. 258-259 (What's in our Solar System?)
- *DK First Space Encyclopedia* pp. 50-51 (The solar system), pp. 82-83 (The asteroid belt)
- *Kingfisher Science Encyclopedia* pp. 398-399 (The Solar System), pg. 413 (Meteors and Meteorites)

Additional Living Books
- *There's No Place Like Space: All About Our Solar System (Cat in the Hat's Learning Library)* by Tish Rabe and Aristides Ruiz
- *Scholastic Reader Level 2: Solar System* by Gregory Vogt
- *Magic School Bus Out of This World : A Book about Space Rocks* by Joanna Cole and Bruce Degen

Notebooking (SCIDAT Logbook Information)

This week, you can have the students complete a night sky journal sheet. You can also have them fill out the logbook sheets for the solar system and asteroids. Here is the information they could include:

Night Sky Journal Sheets
This week, you can look for shooting stars (meteors) when you do your night sky observations.

Astronomy Record Sheets
Solar System
Information Learned
- A solar system includes the sun and everything that orbits around it. This includes the planets, asteroids, moons, comets, and all that space junk.
- In our solar system, the main objects are the eight planets that orbit the sun – Mercury, Venus, Earth, Mars, Jupiter, Saturn, Uranus, and Neptune. There are a few dwarf planets, including Pluto.
- There are also two large asteroid belts, known as the Asteroid Belt and the Kuiper Belt.
- The gravitational pull from the sun keeps all these objects orbiting around it.
- The sun is nearly 1000 times larger than all the planets put together.
- Most of the planets in our solar system have an atmosphere, which is a thin layer of gas that surrounds the planet.
- Our Earth is the only planet in our solar system that is known to have an atmosphere that can currently support life.

Asteroids
Information Learned
- An asteroid is a chunk of iron metal or rock that is orbiting the sun.
- Asteroids vary greatly in size. Some are only meters in length, while some are large enough to be named and considered planetoids.
- There are over 10,000 asteroids that are large enough to be named, such as Ceres, which is about 600 miles wide and is also considered a dwarf planet.
- Eros, one of the named asteroids, had a robot spacecraft landed on in 2001.
- Asteroids have jagged and irregular shapes, so they don't always travel in an even, elliptical pattern as the travel around the sun.
- Most of the asteroids in our solar system orbit the sun in two places—the Asteroid Belt, which is between Mars and Jupiter, and the Kuiper Belt, which is beyond Pluto.
- In 1801, a Sicilian monk named Giuseppe Piazzi discovered the first asteroid in the night sky.
- Some asteroids orbit much closer to Earth; we call those NEAs for Near Earth Asteroids. As they tumble through space, they can be pulled in by Earth's gravity. Once an asteroid enters Earth's atmosphere, it is called a meteor.

Vocabulary
Have the older students look up the following terms in the glossary in the Appendix on pp. 137-138 or in a science encyclopedia. Then, have them copy each definition onto a blank index card or into their SCIDAT logbook.
- **ASTEROID** – A rock orbiting the sun.
- **GRAVITY** – The force that pulls an object towards another larger object.
- **METEOR** – A rock that travels through space and burns up when it enters a planet's atmosphere; also known as a shooting star.
- **SOLAR SYSTEM** – A group of planets and other objects all in orbit around the sun.

Scientific Demonstration: Impact
Materials
- ☑ Marbles
- ☑ Aluminum pan
- ☑ Cornstarch
- ☑ Cocoa Powder
- ☑ Tape measure

Procedure
1. Ahead of time, prepare the planet's surface by pouring a layer of cornstarch on the bottom of the aluminum pan about ½ inch deep and shaking lightly so that the surface is smooth. Then, sprinkle a thin dusting of cocoa powder so that the surface of the cornstarch is mostly covered.
2. Set the pan on the floor and have the students measure 1 foot up from the pan. Have them drop the marble, aiming for the center of the pan.
3. Have the students measure 3 feet up from the pan and have them drop the second marble, aiming for another part of the pan.
4. Remove the marbles, being careful not disturb the holes that were made. Have the students observe the width and depth of the hole created. Ask the students the following:
    - ? What do you notice about the holes that were created?
    - ? How did the two holes differ?

Explanation
The students should see that the marble create an indention on the surface and also displaced some of the cocoa and cornstarch near where it hit. They should also observe that when the marble is dropped from a higher height, the hole formed it a bit deeper and more cornstarch is displaced.

Take it further
Have the students create a work of impact art using cotton balls, paper, and paint. Have the students dip a cotton ball in paint and then drop it on the paper from a height of three feet using the same procedure as in the demonstration.

Multi-Week Projects and Activities

Multi-week Projects
✂ **Solar System Model** – This week, the students will add a basic sun and asteroids to their solar system model. Have the students cut out a large round yellow semi-circle and glue it to the far left of the solar system model for the sun. They will add features to this sun as part of chapter 10. Next have the students add the asteroid belt using pictures of rocks or wadded-up brown paper. This belt should be the following distance from the sun:
- ⇨ Distance for wall version: about 13 in
- ⇨ Distance for lap version: about 7 cm

Activities for This Week
✂ **Solar System Bracelet** – Have the students make a solar system bracelet. You can find directions for this project here:
- 🖱 http://formontana.net/bracelet2.html

✂ **Asteroids** – Have the students make an edible asteroid. You will need ice cream, a plastic baggie, a rolling pin, and cookies, such as vanilla wafers or Oreos. Have them place the cookies in the plastic baggie and crush them with the rolling pin. Then, have them take a scoop of ice cream and roll it around in the crushed cookies. Now that they have made your edible asteroid, the students can eat and enjoy!

✂ **Meteors** – Watch the following video on meteorites with your students:
- 🖱 https://www.youtube.com/watch?v=ZxmuB66iAiQ

Then have the students make their own meteorite. You will need rocks, black paint, and silver glitter. Have them paint their rocks completely black, then dust them with the sliver glitter to make their own meteorite.

Memorization

Copywork/Dictation
☞ **Copywork Sentence**
Our solar system includes the sun, the planets, asteroids, moons, comets, and space junk.

☞ **Dictation Selection**
Gravity pulls an object towards another, larger object. It keeps our planets orbiting around the sun instead of floating off into space. This is the same force that makes objects fall to the ground when you drop them.

Chapter 2 Notes

Chapter 3: Grid Schedule

Supplies Needed

Demo	• Black construction paper, Toothpick, Tape, Flashlight, Large clear bowl, Water
Projects	• Black construction paper, 2 Cotton balls, 1 Small yellow pom-pom, 1 Large red pom-pom, Paint, Markers, 1 Large white sequin

Chapter Summary

The chapter opens with the twins, Summer, and the two animal lab assistants outside the lab gazing up at the stars. The twins learn more about the stars and Summer's spaceship-lab, *Ulysses-1*, which has ties to a company in Switzerland. We then find out that the Man With No Eyebrows did indeed make it to space, where he meets the Triple S agents. We then head back to Summer's to find the twins getting into their IEVA space suits to prepare to launch. Back in space, we find out the Man With No Eyebrows has fired upon the Triple S agents and sent them spinning out into space. Eventually, the come up with a plan to recover – they plan to use the top-secret taser Q-Tip brought. Meanwhile the Man With No Eyebrows is preparing to fire *Thad-1*'s ray gun once more as soon as they twins leave the atmosphere. The chapter wraps us with the twins blasting off just as *Thad-1* is hit with a blast from the Triple-S taser!

Weekly Schedule

	Day 1	Day 2	Day 3	Day 4
Read	☐ Read the section entitled "Smashing Spray-Painted Stars" of Chapter 3 in *SSA Volume 6: Astronomy*.	☐ Read the section entitled "Blast off to the International Space Station" of Chapter 3 in *SSA Volume 6: Astronomy*.	☐ (*Optional*) Read one or all of the assigned pages from the encyclopedia of your choice.	☐ (*Optional*) Read one of the additional books from your library.
Write	☐ Fill out a Astronomy Record Sheet on SL pg. 11 on stars. ☐ Go over the vocabulary words and enter them into the Astronomy Glossary on SL pg. 92.	☐ Fill out a Astronomy Record Sheet on SL pg. 12 on the international space station. ☐ (*Optional*) Add observations to the Night Sky Journal Sheet on SL pg. 8.	☐ (*Optional*) Write narration on the Astronomy Notes Sheet on SL pg. 14. ☐ Add information learned from the demonstration on SL pg. 14.	☐ (*Optional*) Complete the copywork or dictation assignment and add it to the Astronomy Notes sheet on SL pg. 14. ☐ (*Optional*) Fill out the record sheet on SL pg. 16 for one of the projects. ☐ (*Optional*) Take Astronomy Quiz #1.
Do	☐ (*Optional*) Make the Life Cycle of a Star poster.	☐ (*Optional*) Do the International Space Station project.	☐ Do the demonstration entitled "Shining Stars."	☐ Work on the Solar System Model.

Chapter 3: List Schedule

	Supplies Needed
Demo	• Black construction paper, Toothpick, Tape, Flashlight, Large clear bowl, Water
Projects	• Black construction paper, 2 Cotton balls, 1 Small yellow pom-pom, 1 Large red pom-pom, Paint, Markers, 1 Large white sequin

Chapter Summary

The chapter opens with the twins, Summer, and the two animal lab assistants outside the lab gazing up at the stars. The twins learn more about the stars and Summer's spaceship-lab, *Ulysses-1*, which has ties to a company in Switzerland. We then find out that the Man With No Eyebrows did indeed make it to space, where he meets the Triple S agents. We then head back to Summer's to find the twins getting into their IEVA space suits to prepare to launch. Back in space, we find out the Man With No Eyebrows has fired upon the Triple S agents and sent them spinning out into space. Eventually, the come up with a plan to recover – they plan to use the top-secret taser Q-Tip brought. Meanwhile the Man With No Eyebrows is preparing to fire *Thad-1*'s ray gun once more as soon as they twins leave the atmosphere. The chapter wraps us with the twins blasting off just as *Thad-1* is hit with a blast from the Triple-S taser!

Essential To-Do's

Read
☐ Read the section entitled "Smashing Spray-Painted Stars" of Chapter 3 in *SSA Volume 6: Astronomy*.
☐ Read the section entitled "Blast Off to the International Space Station" of Chapter 3 in *SSA Volume 6: Astronomy*.

Write
☐ Fill out a Astronomy Record Sheet on SL pg. 11 on stars.
☐ Go over the vocabulary word and enter it into the Astronomy Glossary on SL pg. 92.
☐ Fill out a Astronomy Record Sheet on SL pg. 12 on the international space station.
☐ Add information learned from the demonstration on SL pg. 14.

Do
☐ Do the demo entitled "Shining Stars."
☐ Work on the Solar System Model.

Optional Extras

Read
☐ Read one or all of the assigned pages from the encyclopedia of your choice.
☐ Read one of the additional books from your library.

Write
☐ Add observations to the Night Sky Journal Sheet on SL pg. 8.
☐ Write a narration on the Astronomy Notes Sheet on SL pg. 14.
☐ Complete the copywork or dictation assignment and add it to the Astronomy Notes sheet on SL pg. 14.
☐ Fill out the record sheet on SL pg. 16 for one of the projects.
☐ Take Astronomy Quiz #1.

Do
☐ Make the Life Cycle of a Star poster.
☐ Do the International Space Station project.

Chapter 3: Zipping Out of this World

Science-Oriented Books

Living Book Spine
- Chapter 3 of *The Sassafras Science Adventures Volume 6: Astronomy*

Optional Encyclopedia Readings
- *Basher Science Astronomy* pg. 20 (International Space Station), pg. 60 (Star Birth Nebula), pg. 80 (White Dwarf)
- *Usborne Children's Encyclopedia* pp. 252-253 (Living in space)
- *DK First Space Encyclopedia* pp. 40-41 (Living in space), pp. 102-103 (A star is born), pp. 104-105 (Death of a star)
- *Kingfisher Science Encyclopedia* pp. 392-392 (Stars), pg. 423 (Space Stations)

Additional Living Books
- *The Sky Is Full of Stars (Let's-Read-and-Find-Out Science 2)* by Franklyn M. Branley and Felicia Bond
- *Jump Into Science: Stars* by Steve Tomecek
- *Stars! Stars! Stars!* by Bob Barner
- *If You Were a Kid Docking at the International Space Station (If You Were a Kid)* by Josh Gregory and Jason Raish
- *International Space Station (Let's-Read-and-Find-Out Science 2)* by Dr. Franklyn M. Branley and True Kelley

Notebooking (SCIDAT Logbook Information)

This week, you can have the students complete a night sky journal sheet. You can also have them fill out the logbook sheets for stars and the international space station. Here is the information they could include:

Night Sky Journal
This week, you can look for different types of stars and for the international space station when you do your night sky observations.

Astronomy Record Sheets
Stars
Information Learned
- A star is a huge ball of exploding gas.
- The stages of a star's life can take millions of years.
- Stars go through a life cycle – they are born, they shine, and one day they die out.
  1. Stars are born in nebulas, which are large, swirling clouds of gas and dust. Inside a nebula clouds of gas clump together and collapse inward, forming the core of the star.
  2. Once the core of the star is formed, it grows hotter and hotter. Eventually, the gas starts to explode and the star begins to shine.
  3. As the gas in the star's core burns out, it begins to die. The star becomes a red giant, meaning that it swells up and turns red.

4. *The gas on the outside burns up and dissipates into space, leaving a small ball known as a white dwarf. As the white dwarf cools, it fades away and the star is gone.*

⇨ *The larger the star, the quicker it burns out; the smaller the star, the star the longer it will shine.*

⇨ *A supernova is a large explosion in space that is produced as a very large star, one much larger than our sun, explodes near the end of its life cycle.*

International Space Station
Information Learned

⇨ *A space station is a man-made structure that is launched into space and orbits the Earth as it orbits the sun.*

⇨ *The International Space Station, also known as the I.S.S., is made from several modules that clip together.*

⇨ *Since 2000, the I.S.S. has been manned by a team of astronaut scientists year-round. The researchers come from several different countries and they perform various experiments to learn how humans and plants are affected by space.*

⇨ *The astronaut scientists live a zero-gravity environment, which means they have to be strapped in when they sleep and exercise.*

⇨ *As of right now, the International Space Station is the most expensive thing man has ever built.*

⇨ *The I.S.S. has solar panels to generate the power it needs and radio antennae and satellite dishes to send signals back to earth with the information from experiments and observations.*

Vocabulary

Have the older students look up the following terms in the glossary in the Appendix on pp. 137-138 or in a science encyclopedia. Then, have them copy each definition onto a blank index card or into their SCIDAT logbook.

- ✎ **Space Station** – A man-made structure that is launched into space and orbits around the sun-orbiting earth.
- ✎ **Star** – A huge ball of exploding gas out in space.
- ✎ **Universe** – The collection of all the matter, space, and energy that exists.

Scientific Demonstration: Shining Stars

Materials
- ☑ Black construction paper
- ☑ Toothpick
- ☑ Tape
- ☑ Flashlight
- ☑ Large clear bowl
- ☑ Water

Procedure

1. Have the students begin by using the toothpick to poke holes in the paper. These are the stars in the night sky. As they create the stars in their night sky, fill the bowl about three-quarters full of water; this will serve as the atmosphere.
2. When the students are done, tape their night sky to the back of the bowl and head into a room without windows. Set the bowl on a flat surface and wait for the water to settle.
3. Turn on the flashlight before turning off the lights in the room. Shine the flashlight on the back of the paper so that the light shines through the star holes and into the atmosphere bowl.

4. Gently tap the bowl so that the water begins to move and let the students observe what happens to the light.

Explanation
The students should see that when you tap the bowl the light moves and their "stars" appear to twinkle. The light is being refracted, or bent, by the water. The same thing happens when light rays from the stars enter Earth's atmosphere, which is why the stars appear to twinkle at night!

Take it Further
Have the students turn on the light in the room and see how this changes the light coming from their stars. (*They should no longer be able to see the distinct "stars" in the bowl. This is because the light in the room is too bright to distinguish the difference. This is why we can't see the stars during the day - the light from the sun is too bright.*)

Multi-Week Projects and Activities

Multi-week Projects
- **Solar System Model** – This week, there is nothing specific to add to the model. However, if you creating the smaller, lap-sized model you can have the students use silver crayon, paint, or glitter to add stars to their model.

Activities For This Week
- **Life Cycle of a Star** – Have the students make a Life Cycle of a Star poster. You will need a sheet of black construction paper, two cotton balls, a small yellow pom-pom, a large red pom-pom, paint, markers, and a large white sequin. Have the students use a pulled-out white cotton ball to make a stellar nebula with its cloud of dust and gas. Then, have them use a small yellow pom-pom for the average star and a large red pom-pom for the red giant. Next, have them paint a cotton ball purple, orange, and a bit of blue. Let it dry and pull it out of shape and use it for the planetary nebula. Finally, have the students use a small white sequin for the white dwarf. You can see what this project looks like on this post:
  - https://elementalscience.com/blogs/science-activities/119870275-the-life-cycle-of-a-star-poster
- **International Space Station** – Have the students watch a live feed from the International Space Station. You can find that here:
  - https://www.youtube.com/watch?v=4993sBLAzGA

We also recommend checking out NASA's YouTube channel. They have several playlists of missions on the International Space Station, simply look for playlists with "I.S.S." in the title.

Memorization

Copywork/Dictation
- **Copywork Sentence**
  A star is a huge ball of exploding gas.
- **Dictation Selection**
  We call everything, all the matter, space, and energy that exists, the universe. Everything in our universe is constantly in motion. Space is full of objects in motion. Some we can see from Earth with just our eyes, like the stars. Some of these objects, we need telescopes to see.

Quiz Information
This week, you can give the students a quiz based on what they learned in chapters 2 and 3. You can find this quiz in the Appendix on pg. 141.

Quiz #1 Answers
1. All of the answers should be circled.
1. Gravity
2. Asteroids
3. Rock
4. Gas
5. 3, 1, 5, 2, 4
6. True
7. Modules

Chapter 3 Notes

Chapter 4: Grid Schedule

	Supplies Needed
Demo	• 2 Thermometers, Cutting board, Clear glass bowl
Projects	• Materials will vary based on the type of planet model you choose to make.

Chapter Summary

The chapter opens with Uncle Cecil losing the twins on the tracking screen, as he expected to happen. We then switch to space, where the twins are with Summer on her spaceship. They learn about how they will travel on the extraterrestrial lines in space, using the zip-zop cuffs on their space suits. And with that, they zip off to the International Space Station, their zip origination point, or zop, for their time in space. They meet their local expert, Yang Bo, and learn about Mercury before heading out on the invisible zip lines to see the planet for themselves. Back out in space, the Man With No Eyebrows recovers *Thad-1* and is back at his mission to stop the twins. Meanwhile, Blaine and Tracey return to the I.S.S. to learn about Venus before heading out into space again to see the planet. The chapter wraps up with Adrianna Archer, who decides to abandon the Man With No Eyebrows and his cause.

Weekly Schedule

	Day 1	**Day 2**	**Day 3**	**Day 4**
Read	☐ Read the section entitled "Mechanical Mercury" of Chapter 4 in *SSA Volume 6: Astronomy*.	☐ Read the section entitled "Venus Vapors" of Chapter 4 in *SSA Volume 6: Astronomy*.	☐ (*Optional*) Read one or all of the assigned pages from the encyclopedia of your choice.	☐ (*Optional*) Read one of the additional books from your library.
Write	☐ Fill out a Astronomy Record Sheet on SL pg. 19 on Mercury. ☐ Go over the vocabulary word and enter it into the Astronomy Glossary on SL pg. 93.	☐ Fill out a Astronomy Record Sheet on SL pg. 20 on Venus. ☐ (*Optional*) Add observations to the Night Sky Journal Sheet on SL pg. 17. ☐ (*Optional*) Write a mission report.	☐ (*Optional*) Write narration on the Astronomy Notes Sheet on SL pg. 23. ☐ Add information learned from the demonstration on SL pg. 23.	☐ (*Optional*) Complete the copywork or dictation assignment and add it to the Astronomy Notes sheet on SL pg. 23. ☐ (*Optional*) Fill out the record sheet on SL pg. 25 for one of the projects.
Do	☐ (*Optional*) Make the Mercury Model.	☐ (*Optional*) Make the Venus Model.	☐ Do the demonstration entitled "Trapped Heat."	☐ Work on the Solar System Model.

Chapter 4: List schedule

	Supplies Needed
Demo	• 2 Thermometers, Cutting board, Clear glass bowl
Projects	• Materials will vary based on the type of planet model you choose to make.

Chapter Summary

The chapter opens with Uncle Cecil losing the twins on the tracking screen, as he expected to happen. We then switch to space, where the twins are with Summer on her spaceship. They learn about how they will travel on the extraterrestrial lines in space, using the zip-zop cuffs on their space suits. And with that, they zip off to the International Space Station, their zip origination point, or zop, for their time in space. They meet their local expert, Yang Bo, and learn about Mercury before heading out on the invisible zip lines to see the planet for themselves. Back out in space, the Man With No Eyebrows recovers *Thad-1* and is back at his mission to stop the twins. Meanwhile, Blaine and Tracey return to the I.S.S. to learn about Venus before heading out into space again to see the planet. The chapter wraps up with Adrianna Archer, who decides to abandon the Man With No Eyebrows and his cause.

Essential To-Do's

Read
☐ Read the section entitled "Mechanical Mercury" of Chapter 4 in *SSA Volume 6: Astronomy*.
☐ Read the section entitled "Venus Vapors" of Chapter 4 in *SSA Volume 6: Astronomy*.

Write
☐ Fill out a Astronomy Record Sheet on SL pg. 19 on Mercury.
☐ Go over the vocabulary word and enter it into the Astronomy Glossary on SL pg. 93.
☐ Fill out a Astronomy Record Sheet on SL pg. 20 on Venus.
☐ Add information learned from the demonstration on SL pg. 23.

Do
☐ Do the demonstration entitled "Trapped Heat."
☐ Work on the Solar System Model.

Optional Extras

Read
☐ Read one or all of the assigned pages from the encyclopedia of your choice.
☐ Read one of the additional books from your library.

Write
☐ Add observations to the Night Sky Journal Sheet on SL pg. 17.
☐ Write a narration on the Astronomy Notes Sheet on SL pg. 23.
☐ Complete the copywork or dictation assignment and add it to the Astronomy Notes sheet on SL pg. 23.
☐ Fill out the record sheet on SL pg. 25 for one of the projects.
☐ Write a mission report.

Do
☐ Make the Mercury Model.
☐ Make the Venus Model.

Chapter 4: Space Sandwiches and Planetary Potato Chips

Science-Oriented Books

Living Book Spine
- Chapter 4 of *The Sassafras Science Adventures Volume 6: Astronomy*

Optional Encyclopedia Readings
- *Basher Science Astronomy* pg. 12 (Mercury), pg. 14 (Venus)
- *Usborne Children's Encyclopedia* pg. 264 (Mercury), pg. 265 (Venus)
- *DK First Space Encyclopedia* pp. 56-57 (Mercury), pp. 58-59 (The Morning Star)
- *Kingfisher Science Encyclopedia* pg. 403 (Mercury), pg. 404 (Venus)

Additional Living Books
- *Mercury (First Facts: Solar System)* by Adele Richardson
- *Mercury (Scholastic News Nonfiction Readers: Space Science)* by Christine Taylor-Butler
- *Mercury and Venus (Up in Space)* by Rosalind Mist
- *Venus (True Books: Space)* by Elaine Landau
- *Venus (First Facts: Solar System)* by Adele Richardson
- *Venus (Scholastic News Nonfiction Readers: Space Science)* by Melanie Chrismer

Notebooking (SCIDAT Logbook Information)

This week, you can have the students complete a night sky journal sheet. You can also have them fill out the logbook sheets for Mercury and Venus. Here is the information they could include:

Night Sky Journal
This week, you can look for Mercury and Venus when you do your night sky observations. The following article has information about how to spot the planets in the night sky:
- https://www.space.com/39240-when-to-see-planets-in-the-sky.html (NOTE—*This article link contains information for spotting the planets in 2019. It will still help with spotting the planets at any time, but you can check the resource page in the introduction of this guide for an updated link.*)

Astronomy Record Sheets
Mercury
Information Learned
- Mercury is the closest planet to the sun.
- It is relatively small compared to the other planets in our solar system, and it is very, very hot.
- During the day it gets up to over 800 degrees Fahrenheit. However, the temperature ranges on the surface of Mercury are the greatest of all the planets, so at night it can dip down to freezing temperatures that can nearly reach -300 degrees Fahrenheit.

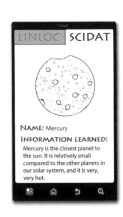

⇨ *Mercury is named after the Roman messenger god.*
⇨ *The planet is one-third the size of Earth.*
⇨ *It is almost as heavy as the earth due to the fact that its core is made of very dense metal.*
⇨ *The first glimpse of the surface of Mercury was caught in 1973 from a space probe named Mariner-10.*
⇨ *It's a surface that is covered with numerous craters; the largest of these is called Caloris Basin, and it is nearly 800 miles across.*
⇨ *A year on Mercury takes only about 88 Earth days and a day lasts over 58 Earth days.*

Venus
Information Learned

⇨ *Venus is the second planet from the sun in our solar system.*
⇨ *It is a small, hot planet, even hotter than Mercury.*
⇨ *Venus is covered with a thick layer of carbon dioxide gas which traps heat, making it extremely hot.*
⇨ *After the moon, Venus is the brightest object in our night sky. It appears like a bright star in both the early morning and early evening.*
⇨ *The surface of Venus is not visible through its thick and hazy atmosphere. However, we do have pictures of the planet's surface, thanks to a series of satellites called Venera' that were sent out by the Soviet Union. The United States also sent the orbiter Pioneer and the probe Magellan, to travel around Venus and use radar to map out the surface. We now know that the surface of Venus is rocky and is covered with mountains, canyons, shallow craters, volcanoes, and hardened lava flows.*
⇨ *A year on Venus takes nearly 225 earth days, but a day takes over 243 earth days.*

Vocabulary

Have the older students look up the following term in the glossary in the Appendix on pp. 137-138 or in a science encyclopedia. Then, have them copy the definition onto a blank index card or into their SCIDAT logbook.

✎ **PLANET** – A large ball of rock or gas that travels around a star.

Scientific Demonstration: Trapped Heat

Materials
☑ 2 Thermometers
☑ Cutting board
☑ Clear glass bowl

Procedure
1. Find a spot that gets direct sunlight and set down the cutting board. Have the students place the two thermometers on the board.
2. Have the students observe and record the temperature of the two thermometers on a sheet of paper. Then, cover one of the thermometers with the upside-down glass bowl.
3. Wait 5 to 10 minutes before having the students record the temperature of the two thermometers once more. After they are done, pick up the board and move the thermometers to the inside of a dark closet.
4. Wait 5 to 10 more minutes before having the students record the temperature of the two thermometers again. Ask the students the following:
   ❓ What happened to the temperature on the thermometer that was left in the open?
   ❓ What happened to the temperature on the thermometer that was under the bowl?

The Sassafras Guide to Astronomy ~ Chapter 4

Explanation
The students should see that the temperature of the thermometer in the bowl did not increase or decrease as much as the thermometer left out in the open.

Take it Further
Repeat the experiment, but this time put a sheet of black plastic under the second thermometer before covering it with a bowl. (*The students should see that the heat rises higher and holds for longer than before.*)

Multi-Week Projects and Activities

Multi-week Projects
- **Solar System Model** – This week, the students will add the first two planets, Mercury and Venus, to their solar system model. Have the students cut our the planets from copy paper or construction paper using the diameters below. Have them color the planets using a picture from the suggested books or from the Internet. When they are done, have them add the planets to their solar system using the distances below.

  Mercury
  - Diameter (wall version): 3/4 in
  - Distance (wall version): 2 in
  - Diameter (lap version): 0.4 cm
  - Distance (lap version): 1 cm

  Venus
  - Diameter (wall version): 1-3/4 in
  - Distance (wall version): 3 in
  - Diameter (lap version): 0.9 cm
  - Distance (lap version): 1.5 cm

Activities For This Week
- **Mercury Model** – Have the students make a 3D model of Mercury. You can have them paint a Styrofoam ball, or you can have them make a paper-mâché model. (**NOTE**—*See the directions found on Appendix pg. 129 for the paper-mâché planet.*)
- **Venus Model** – Have the students make a 3D model of Venus. You can have them paint a Styrofoam ball, or you can have them make a paper-mâché model. (**NOTE**—*See the directions found on Appendix pg. 129 for the paper-mâché planet.*)
- **Mission Report** – Have the students research and write a brief report about one of the missions to either Mercury or Venus. They can choose from the two space missions to Mercury: *Mariner 10* or *MESSENGER*, or from the two space missions to Venus: *Magellan* or *Venus Express*. This report should include the dates of the mission, several pictures, and a summary of what was found.

Memorization

Copywork/Dictation
☞ **Copywork Sentence**
   Mercury is the closest planet to the sun. Venus is the second planet from the sun.

☞ **Dictation Selection**
   A year on a planet is the time it takes for the planet to orbit the sun. A day on a planet is the time it takes for a planet to completely rotate, resulting in both a sunrise and sunset. A year on Mercury takes only about 88 Earth days and a day lasts over 58 Earth days. A year on Venus takes nearly 225 Earth days, but a day takes over 243 Earth days.

Chapter 4 Notes

Chapter 5: Grid Schedule

	Supplies Needed
Demo	• A solar-powered mini-car kit (or a DC motor, Solar panel with wires, 2 Sets of wheels with axles or steel wires, Cardboard, Hot glue, Plastic tubing)
Projects	• Materials will vary based on the type of planet model you choose to make.

Chapter Summary
The chapter opens with the Sassafras twins floating out in front of Venus. They soon return to the I.S.S. where they meet the rest of the astronauts. Captain Diana Sturgess tells them about Earth and as she is wrapping up, Sanders runs in and announces that it was sabotage. The twins learn about Brett Frye and his unexpected launch in the Mars Module. The captain asks for Blaine and Tracey's help in getting to the bottom of the problem. The Sassafras twins interview the crew members and see that, with the exception of the captain and their local expert, everyone has a motive. During Yang Bo's interview, he tells the twins about Mars and suggests that they attempt to rescue Brett when they go to see Mars. The twins agree to that mission. The chapter wraps up with us finding out that sure enough, the circuit board was sabotaged – with the queen from a magnetic chess board!

Weekly Schedule

	Day 1	Day 2	Day 3	Day 4
Read	☐ Read the section entitled "Earth in Space" of Chapter 5 in *SSA Volume 6: Astronomy*.	☐ Read the section entitled "Mars Module" of Chapter 5 in *SSA Volume 6: Astronomy*.	☐ *(Optional)* Read one or all of the assigned pages from the encyclopedia of your choice.	☐ *(Optional)* Read one of the additional books from your library.
Write	☐ Fill out a Astronomy Record Sheet on SL pg. 22 on Earth. ☐ Go over the vocabulary words and enter them into the Astronomy Glossary on SL pg. 93.	☐ Fill out a Astronomy Record Sheet on SL pg. 23 on Mars. ☐ *(Optional)* Add observations to the Night Sky Journal Sheet on SL pg. 18. ☐ *(Optional)* Write a mission report.	☐ *(Optional)* Write narration on the Astronomy Notes Sheet on SL pg. 24. ☐ Add information learned from the demonstration on SL pg. 24.	☐ *(Optional)* Complete the copywork or dictation assignment and add it to the Astronomy Notes sheet on SL pg. 24. ☐ *(Optional)* Fill out the record sheet on SL pg. 26 for one of the projects. ☐ *(Optional)* Take Astronomy Quiz #2.
Do	☐ *(Optional)* Make the Earth Model. ☐ *(Optional)* Do the Venus vs. Earth Venn Diagram.	☐ *(Optional)* Make the Mars model.	☐ Do the demonstration entitled "Solar Rover."	☐ Work on the Solar System Model.

Chapter 5: List schedule

	Supplies Needed
Demo	• A solar-powered mini-car kit (or a DC motor, Solar panel with wires, 2 Sets of wheels with axles or steel wires, Cardboard, Hot glue, Plastic tubing)
Projects	• Materials will vary based on the type of planet model you choose to make.

Chapter Summary

The chapter opens with the Sassafras twins floating out in front of Venus. They soon return to the I.S.S. where they meet the rest of the astronauts. Captain Diana Sturgess tells them about Earth and as she is wrapping up Sanders runs in and announces that it was sabotage. The twins learn about Brett Frye and his unexpected launch in the Mars Module. The captain asks for Blaine's and Tracey's help in getting to the bottom of the problem. The Sassafras twins interview the crew members and see that with the exception of the captain and their local expert, everyone has a motive. During Yang Bo's interview, he tells the twins about Mars and then suggests that they attempt to rescue Brett when they go to see Mars. The twins agree to that mission. The chapter wraps up with us finding out that sure enough, the circuit board was sabotaged – with the queen from a magnetic chess board!

Essential To-Do's

Read
☐ Read the section entitled "Earth in Space" of Chapter 5 in *SSA Volume 6: Astronomy*.
☐ Read the section entitled "Mars Module" of Chapter 5 in *SSA Volume 6: Astronomy*.

Write
☐ Fill out a Astronomy Record Sheet on SL pg. 22 on Earth.
☐ Go over the vocabulary words and enter them into the Astronomy Glossary on SL pg. 93.
☐ Fill out a Astronomy Record Sheet on SL pg. 23 on Mars.
☐ Add information learned from the demonstration on SL pg. 24.

Do
☐ Do the demo entitled "Solar Rover."
☐ Work on the Solar System Model.

Optional Extras

Read
☐ Read one or all of the assigned pages from the encyclopedia of your choice.
☐ Read one of the additional books from your library.

Write
☐ Add observations to the Night Sky Journal Sheet on SL pg. 24.
☐ Write a narration on the Astronomy Notes Sheet on SL pg. 24.
☐ Complete the copywork or dictation assignment and add it to the Astronomy Notes sheet on SL pg. 24.
☐ Fill out the record sheet on SL pg. 26 for one of the projects.
☐ Write a mission report.
☐ Take Astronomy Quiz #2.

Do
☐ Make the Earth and Mars models.
☐ Do the Venus vs. Earth Venn Diagram.

Chapter 5: The Rest of the Inner Planetary Scoop

Science-Oriented Books

Living Book Spine
- Chapter 5 of *The Sassafras Science Adventures Volume 6: Astronomy*

Optional Encyclopedia Readings
- *Basher Science Astronomy* pg. 16 (Earth), pg. 22 (The Moon), pg. 24 (Mars)
- *Usborne Children's Encyclopedia* pp. 8-9 (Our Planet), pp. 260-261 (The Moon), pp. 266-267 (Mars)
- *DK First Space Encyclopedia* pp. 60-61 (Third rock from the Sun), pg. 62-63 (The moon), pp. 64-65 (The red planet)
- *Kingfisher Science Encyclopedia* pg. 400 (Earth and the Moon - Earth Section only), pg. 405 (Mars)

Additional Living Books
- *Planet Earth/Inside Out* by Gail Gibbons
- *The Magic School Bus Inside the Earth (Magic School Bus)* by Joanna Cole and Bruce Degen
- *See Inside Planet Earth (Usborne Flap Book)* by Katie Daynes and Peter Allen
- *You Are the First Kid on Mars* by Patrick O'Brien
- *Mission to Mars (Let's-Read-and-Find-Out Science 2)* by Franklyn M. Branley and True Kelley
- *Mars (Scholastic News Nonfiction Readers: Space Science)* by Melanie Chrismer

Notebooking (SCIDAT Logbook Information)

This week, you can have the students complete a night sky journal sheet. You can also have them fill out the logbook sheets for Earth and Mars. Here is the information they could include:

Night Sky Journal
This week, you can look for Mars when you do your night sky observations. The following article has information about how to spot the planets in the night sky:
- https://www.space.com/39240-when-to-see-planets-in-the-sky.html (NOTE—*This article link contains information for spotting the planets in 2019. It will still help with spotting the planets at any time, but you can check the resource page in the introduction of this guide for an updated link.*)

Astronomy Record Sheets
Earth
Information Learned

- Earth is the third planet from the sun.
- Earth is the only planet in our solar system that we know of that can support life.
- The atmosphere on Earth is a unique combination of oxygen, nitrogen, and other important elements.
- Earth's atmosphere extends out six hundred miles from the surface, protecting people from the harmful rays of the sun.
- The presence of water in liquid form, along with Earth's dynamic weather patterns, help to provide the conditions needed for Earth's diverse plant and animal life.

⇨ *It takes precisely 365.26 days for Earth to fully orbit around the sun.*
⇨ *A full rotation of the Earth, or rather a day, takes 23 hours and 56 minutes.*

Mars
Information Learned

⇨ *Mars is the fourth planet from the sun in our solar system.*
⇨ *The surface of the planet looks like a great big desert of red, iron-rich dust with volcanoes and canyons dotting the landscape.*
⇨ *The biggest volcano on Mars is called Mount Olympus, and it is the largest known volcano in the entire solar system.*
⇨ *The largest canyon on Mars, Valles Marineris, could spread across the whole United States.*
⇨ *Mars has storms, clouds, and fog, similar to what we have on earth, but it's much colder than our home planet.*
⇨ *There is a thin atmosphere of poison gas covering the planet, and each of its poles are covered with frozen ice made from a mix of carbon dioxide and water.*
⇨ *Mars is half the size of the Earth and is the closest planet to Earth.*
⇨ *Mars was first visited in 1976 by the space probe Viking-1. But most of the photos that we have today of the surface of Mars were taken by Pathfinder, a probe that was sent to the planet in the late 1990s.*
⇨ *A day on Mars takes around 24 hours and 37 minutes. A year on Mars takes 687 earth days.*

Vocabulary

Have the older students look up the following terms in the glossary in the Appendix on pp. 137-138 or in a science encyclopedia. Then, have them copy the definition onto a blank index card or into their SCIDAT logbook.

↷ ATMOSPHERE – A layer of gas that surrounds a planet.
↷ ORBIT – The path of an object in space.

Scientific Demonstration: Solar Rover

Materials
☑ A solar-powered mini-car kit
☑ OR for the DIY version, you will need a DC motor, a solar panel with wires, 2 sets of wheels with axles or steel wires to make axles, cardboard, hot glue, and a few inches of plastic tubing.

Procedure
1. Have the students follow the directions provided in the kit or if you want to DIY your solar rover, follow the directions found here:
   https://www.instructables.com/id/How-to-Make-Solar-Car-DIY-Mini-Car/
   (NOTE – *The solar car kit included in the Sassafras Year 3 Experiment Kit requires full sun or a very strong flashlight to work. If you find that your car is not moving, it may be because the light is not strong enough to charge the solar cell. You can take the car outside where it will be in direct sunlight or aim a strong flashlight directly on the solar cell to power-up the car.*)

Multi-Week Projects and Activities

Multi-week Projects
✂ SOLAR SYSTEM MODEL – This week, the students will add the next two planets, Earth and Mars, to their solar system model. Have the students cut our the planets from copy paper or construction

paper using the diameters below. Have them color the planets using a picture from the suggested books or from the Internet. When they are done, have them add the planets to their solar system using the distances below.

Earth
- Diameter (wall version): 2 in
- Distance (wall version): 4 in
- Diameter (lap version): 1 cm
- Distance (lap version): 2 cm

Mars
- Diameter (wall version): 1-1/8 in
- Distance (wall version): 6 in
- Diameter (lap version): 0.6 cm
- Distance (lap version): 3 cm

Activities For This Week

- **Earth Model** – Have the students make a 3D model of Earth You can have them paint a Styrofoam ball, or you can have them make a paper-mâché model. (**NOTE**—*See the directions found on Appendix pg. 129 for the paper-mâché planet.*)
- **Venus vs. Earth** – Have the students compare Venus and Earth using the Venn Diagram in Appendix on pg. 130. For example, you could include:
  - Both are inner planets and both have thick atmospheres.
  - Venus is very hot, has a toxic atmosphere, and has lots of volcanoes.
  - Earth has water and oxygen, supports life, and is farther from the sun.
- **Mars Model** – Have the students make a 3D model of Mars. You can have them paint a Styrofoam ball, or you can have them make a paper-mâché model. (**NOTE**—*See the directions found on Appendix pg. 129 for the paper-mâché planet.*)
- **Mission Report** – Have the students research and write a brief report about one of the missions to either the moon or Mars. They can choose one of the most famous journeys to the moon, *Apollo 11*, or from one of the space missions to Mars: *Mariner, Pathfinder, Sojourner,* or *Curiosity*. This report should include the dates of the mission, several pictures, and a summary of what was found.

Memorization

Copywork/Dictation

☞ **Copywork Sentence**

Our planet, Earth, is the third from the sun. Mars is known as the red planet.

☞ **Dictation Selection**

Earth is the only planet in our solar system that we know of that can support life. The atmosphere on Earth is a unique combination of oxygen, nitrogen, and other important elements. Mars is the fourth planet from the sun in our solar system. The surface of the planet looks like a great big desert of red, iron-rich dust with volcanoes and canyons dotting the landscape.

Quiz Information

This week, you can give the students a quiz based on what they learned in chapters 4 and 5. You can find this quiz in the Appendix on pg. 143.

Quiz #2 Answers

1. Mercury, Venus, Earth, Mars
2. One-third of the
3. Carbon Dioxide
4. False (*Mercury is the closest planet to the sun.*)
5. Life
6. Mars
7. True

Chapter 5 Notes

Chapter 6: Grid Schedule

Supplies Needed

Demo	• Bowl, Milk, Food Coloring, Liquid soap, Toothpick
Projects	• Materials will vary based on the type of planet model you choose to make.

Chapter Summary

The chapter opens with Blaine, Tracey, REESE, and SLIM heading off to Mars in hopes of rescuing Brett. The Man With No Eyebrows overhears their mission and vows to stop them. The twins and the robots zip off, but something happens mid-zip and they end up changing their course for Jupiter. After they learn a bit about the planet, they attempt to zip to Mars once more. They make it to the red planet and find Brett alive, but in need of help. They get ready to zip back to the I.S.S. and realize they are missing the extra cuff. They soon find out that the Man With No Eyebrows in the Dark Cape has also made it to Mars and he has the cuff, using it to whisk Brett away to Saturn. The twins and the robots follow, but they just miss the duo. SLIM and REESE share about the planet with Blaine and Tracey. The chapter wraps up with them spotting the Man With No Eyebrows and Brett just before they disappear once more out into space.

Weekly Schedule

	Day 1	Day 2	Day 3	Day 4
Read	☐ Read the section entitled "Jumping Jupiter" of Chapter 6 in *SSA Volume 6: Astronomy*.	☐ Read the section entitled "Searching Saturn" of Chapter 6 in *SSA Volume 6: Astronomy*.	☐ (*Optional*) Read one or all of the assigned pages from the encyclopedia of your choice.	☐ (*Optional*) Read one of the additional books from your library.
Write	☐ Fill out a Astronomy Record Sheet on SL pg. 29 on Jupiter. ☐ Go over the vocabulary word and enter it into the Astronomy Glossary on SL pg. 93.	☐ Fill out a Astronomy Record Sheet on SL pg. 30 on Saturn. ☐ (*Optional*) Add observations to the Night Sky Journal Sheet on SL pg. ___. ☐ (*Optional*) Write a mission report.	☐ (*Optional*) Write narration on the Astronomy Notes Sheet on SL pg. 33. ☐ Add information learned from the demonstration on SL pg. 33.	☐ (*Optional*) Complete the copywork or dictation assignment and add it to the Astronomy Notes sheet on SL pg. 33. ☐ (*Optional*) Fill out the record sheet on SL pg. 35 for one of the projects.
Do	☐ (*Optional*) Make the Jupiter model.	☐ (*Optional*) Make the Saturn model.	☐ Do the demonstration entitled "Stormy Swirls."	☐ Work on the Solar System Model.

Chapter 6: List Schedule

	Supplies Needed
Demo	• Bowl, Milk, Food Coloring, Liquid soap, Toothpick
Projects	• Materials will vary based on the type of planet model you choose to make.

Chapter Summary

The chapter opens with Blaine, Tracey, REESE, and SLIM heading off to Mars in hopes of rescuing Brett. The Man With No Eyebrows overhears their mission and vows to stop them. The twins and the robots zip off, but something happens mid-zip and they end up changing their course for Jupiter. After they learn a bit about the planet, they attempt to zip to Mars once more. They make it to the red planet and find Brett alive, but in need of help. They get ready to zip back to the I.S.S. and realize they are missing the extra cuff. They soon find out that the Man With No Eyebrows in the Dark Cape has also made it to Mars and he has the cuff, using it to whisk Brett away to Saturn. The twins and the robots follow, but they just miss the duo. SLIM and REESE share about the planet with Blaine and Tracey. The chapter wraps up with them spotting the Man With No Eyebrows and Brett just before they disappear once more out into space.

Essential To-Do's

Read
- ☐ Read the section entitled "Jumping Jupiter" of Chapter 6 in *SSA Volume 6: Astronomy*.
- ☐ Read the section entitled "Searching Saturn" of Chapter 6 in *SSA Volume 6: Astronomy*.

Write
- ☐ Fill out a Astronomy Record Sheet on SL pg. 29 on Jupiter.
- ☐ Go over the vocabulary word and enter it into the Astronomy Glossary on SL pg. 93.
- ☐ Fill out a Astronomy Record Sheet on SL pg. 30 on Saturn.
- ☐ Add information learned from the demonstration on SL pg. 33.

Do
- ☐ Do the demonstration entitled "Stormy Swirls."
- ☐ Work on the Solar System Model.

Optional Extras

Read
- ☐ Read one or all of the assigned pages from the encyclopedia of your choice.
- ☐ Read one of the additional books from your library.

Write
- ☐ Add observations to the Night Sky Journal Sheet on SL pg. 33.
- ☐ Write a narration on the Astronomy Notes Sheet on SL pg. 33.
- ☐ Complete the copywork or dictation assignment and add it to the Astronomy Notes sheet on SL pg. 33.
- ☐ Fill out the record sheet on SL pg. 35 for one of the projects.
- ☐ Write a mission report.

Do
- ☐ Make the Jupiter model.
- ☐ Make the Saturn model.

Chapter 6: The Red Planet Rescue Mission

Science-Oriented Books

Living Book Spine
- Chapter 6 of *The Sassafras Science Adventures Volume 6: Astronomy*

Optional Encyclopedia Readings
- *Basher Science Astronomy* pg. 32 (Jupiter), pg. 34 (Jupiter's moons), pg. 36 (Saturn)
- *Usborne Children's Encyclopedia* pp. 268-269 (Jupiter and Saturn)
- *DK First Space Encyclopedia* pp. 66-67 (King of planets), pp. 68-69 (Jupiter's moons), pp. 70-71 (Saturn)
- *Kingfisher Science Encyclopedia* pg. 406 (Jupiter), pg. 407 (Saturn)

Additional Living Books
- *Destination: Jupiter* by Seymour Simon
- *Jupiter: The Largest Planet (Our Solar System)* by Daisy Allyn
- *Planet Jupiter (True Books)* by Ann O. Squire
- *Jupiter (Scholastic News Nonfiction Readers: Space Science)* by Christine Taylor-Butler
- *Jupiter (Blastoff! Readers: Exploring Space)* by Derek Zobel
- *Saturn* by Seymour Simon
- *Saturn (True Books)* by Elaine Landau
- *Saturn: The Ringed Planet (Our Solar System)* by Daisy Allyn
- *Planet Saturn (True Books)* by Ann O. Squire
- *Saturn (Scholastic News Nonfiction Readers: Space Science)* by Christine Taylor-Butler
- *Jupiter and Saturn (Up in Space)* by Rosalind Mist

Notebooking (SCIDAT Logbook Information)

This week, you can have the students complete a night sky journal sheet. You can also have them fill out the logbook sheets for Jupiter and Saturn. Here is the information they could include:

Night Sky Journal
This week, you can look for Jupiter and Saturn when you do your night sky observations. The following article has information about how to spot the planets in the night sky:
- https://www.space.com/39240-when-to-see-planets-in-the-sky.html (**NOTE**—*This article link contains information for spotting the planets in 2019. It will still help with spotting the planets at any time, but you can check the resource page in the introduction of this guide for an updated link.*)

Astronomy Record Sheets
Jupiter
Information Learned
- ⇨ Jupiter is the largest planet in our solar system and can easily be spotted with the naked eye from Earth.
- ⇨ It is the fourth brightest object in the night sky and is sometimes referred to as *King of Planets*.
- ⇨ Jupiter is made mainly of gas, which is why we call it one of the gas giants.

⇨ *Jupiter mainly has hydrogen gas, but there is some helium as well. Below its very thick layer of gas, there is a layer of liquid hydrogen surrounding an iron core.*
⇨ *The atmosphere covering Jupiter has both dark and light clouds. Scientists have divided these strips of clouds into different bands and zones. The wind flows in one direction in the zones, and in the opposite direction in the bands.*
⇨ *Jupiter has an atmosphere that is also extremely stormy.*
⇨ *One of the largest of these storms is known as the Great Red Spot. This spot has been visible for over 400 years, and it generally looks like a large red oval. It changes in size and shape from year to year. Overall the Great Red Pot is three times the size of Earth.*
⇨ *Jupiter has more than fifty moons. These moons vary in size from about six miles across to quite a bit larger.*
⇨ *The four largest of Jupiter's moons, Ganymede, Callisto, Io, and Europa, can be seen from Earth with a telescope.*
⇨ *A year on Jupiter takes about 12 Earth years*
⇨ *A day takes about 10 Earth hours.*

Saturn
Information Learned

⇨ *Saturn is one of the outer planets. It is the sixth planet from the sun.*
⇨ *Saturn is the second largest planet in our solar system.*
⇨ *It is light for its size since it is composed mainly of gas. In fact, Saturn is so light, it could float on the ocean if there were one big enough to hold it.*
⇨ *Saturn's rings are made up of ice, rocks, and dust that orbit the planet. The ice rings reflect light, which is why Saturn's rings are so visible.*
⇨ *Saturn is largely considered to be the most beautiful planet in our solar system.*
⇨ *The atmosphere of Saturn is made up of ammonia, water, and methane that is colored by phosphorus and other elements. This gives Saturn its very colorful shades of cream, yellow, and brown.*
⇨ *The rest of Saturn is composed of liquid hydrogen with a solid iron core, just like Jupiter. Also, like Jupiter, Saturn has frequent violent storms.*
⇨ *Saturn has over 50 confirmed moons that are mostly made of ice, except for Titan, which is Saturn's largest moon.*
⇨ *A year on Saturn takes nearly 30 Earth years.*
⇨ *A day takes only about 11 Earth hours.*

Vocabulary

Have the older students look up the following term in the glossary in the Appendix on pp. 137-138 or in a science encyclopedia. Then, have them copy each definition onto a blank index card or into their SCIDAT logbook.

↻ MOON – A natural satellite in orbit around a planet.

Scientific Demonstration: Stormy Swirls

Materials
☑ Bowl
☑ Milk
☑ Food Coloring

☑ Liquid soap
☑ Toothpick

Procedure
1. Have the students pour a thin layer of warm milk into a bowl for the students.
2. Next have the students add a drop of each color (blue, green, red, and yellow) of food coloring in four different places in the bowl.
3. Then, have the students add a drop of liquid dish soap to the center of the bowl.
4. Have the students observe the bowl as you ask them, "What do you notice is happening to the colors in the bowl?"

Explanation
The students should see the colors swirl and mix around in the bowl after the dish soap touches the surface. In the bowl, the dish soap breaks the surface tension on the milk, causing the milk molecules to tumble and mix all around the bowl. As the molecules mix, they pick up the food coloring molecules, creating storm swirls of colorful liquid throughout the bowl. Unlike in this demonstration, scientists don't exactly know why the gases on Jupiter swirl and why the Great Red Spot is there. There are lots of good ideas, but since Jupiter doesn't have a calm, solid surface, it is very difficult for us to study the planet.

Take it further
Have the students make a swirl-art Jupiter. You will need liquid starch, paper, a pie pan, a craft stick, and red, orange, yellow, gray, or brown acrylic paint. Have the students trace the pie pan on the paper. Then have them cut out a circle that is slightly smaller so that it will fit into the pan. While they cut out the circle, pour enough liquid starch into the pie pan to cover the bottom and add several drops of the paint in different place in the starch. Have the students use the craft stick to swirl the paint drops and then gently set the paper on top of the swirls. Peel back the paper and let it dry before gluing it on a black sheet of construction paper.

Multi-Week Projects and Activities

Multi-week Projects

✂ **Solar System Model** – This week, the students will add the next two planets, Jupiter and Saturn, to their solar system model. Have the students cut our the planets from copy paper or construction paper using the diameters below. Have them color the planets using a picture from the suggested books or from the Internet. When they are done, have them add the planets to their solar system using the distances below.

Jupiter
⇨ Diameter (wall version): 22 in
⇨ Distance (wall version): 1 foot, 9 in
⇨ Diameter (lap version): 10.1 cm
⇨ Distance (lap version): 10.6 cm

Saturn
⇨ Diameter (wall version): 20 in
⇨ Distance (wall version): 3 feet, 2 in
⇨ Diameter (lap version): 10 cm
⇨ Distance (lap version): 19 cm

Activities For This Week

✂ **Jupiter Model** – Have the students make a 3D model of Jupiter You can have them paint a Styrofoam ball, or you can have them make a paper-mâché model. (**Note**—*See the directions found on Appendix pg. 129 for the paper-mâché planet.*)

✂ **Saturn Model** – Have the students make a 3D model of Saturn. You can have them paint a Styrofoam ball, or you can have them make a paper-mâché model. (**Note**—*See the directions found on Appendix pg. 129 for the paper-mâché planet.*)

✂ MISSION REPORT – Have the students research and write a brief report about one of the missions to either the Jupiter or Saturn. They can choose the *Galileo* space mission to Jupiter or the *Cassini* space mission to Saturn. This report should include the dates of the mission, several pictures, and a summary of what was found.

MEMORIZATION

COPYWORK/DICTATION

☞ **COPYWORK SENTENCE**

Jupiter is the largest planet in our solar system. Saturn has rings.

☞ **DICTATION SELECTION**

Jupiter is the largest planet in our solar system. It is the fourth brightest object in the night sky and is sometimes referred to as King of Planets. Jupiter is made mainly of gas, which is why we call it one of the gas giants. Saturn is also a gas giant. The planet's rings are made up of ice, rocks, and dust that orbit the planet. The ice rings reflect light, which is why Saturn's rings are so visible.

CHAPTER 6 NOTES

Chapter 7: Grid Schedule

	Supplies Needed
Demo	• Marble, Smooth pie plate or cake pan
Projects	• Materials will vary based on the type of planet model you choose to make.

Chapter Summary

The chapter opens with the two Triple S agents, Evan DeBlose and Q-Tip finally reaching the targeted satellite. It then quickly switches to Adrianna Archer, the turncoat, who is now sitting outside of Uncle Cecil's house, watching his every move. Back in space, Blaine and Tracey decide to continue the search for Brett and the Dark Cape by heading out to the next planet, Uranus. After learning about the planet and realizing the duo they seek is not there, they head onto Neptune. Meanwhile on the I.S.S., the struggle for justice continues and Partha is one step closer to being kicked off the space station. Back at Neptune, the twins learn about the planet, but their focus remains on finding Brett. When they don't see him, they decide to return to the I.S.S., defeated. But close by the Man With No Eyebrows in the Dark Cape has a change of heart and debates about where to take Brett so that he will be seen and rescued. Back at the I.S.S., the twins share their failure with Yang Bo, but as they are speaking, Brett is spotted just outside the receiving room. They bring him onto the space station, and we switch to the Man With No Eyebrows where we find out that he has decided to now be a force for good. The chapter wraps up on the I.S.S. where Brett wakes up and everyone is waiting with bated breath to hear his side of the story.

Weekly Schedule

	Day 1	Day 2	Day 3	Day 4
Read	☐ Read the section entitled "Unsurmountable Uranus" of Chapter 7 in *SSA Volume 6: Astronomy*.	☐ Read the section entitled "Nearing Neptune" of Chapter 7 in *SSA Volume 6: Astronomy*.	☐ (*Optional*) Read one or all of the assigned pages from the encyclopedia of your choice.	☐ (*Optional*) Read one of the additional books from your library.
Write	☐ Fill out a Astronomy Record Sheet on SL pg. 31 on Uranus. ☐ Go over the vocabulary word and enter it into the Astronomy Glossary on SL pg. 94.	☐ Fill out a Astronomy Record Sheet on SL pg. 32 on Neptune. ☐ (*Optional*) Add observations to the Night Sky Journal Sheet on SL pg. 28. ☐ (*Optional*) Write a mission report.	☐ (*Optional*) Write narration on the Astronomy Notes Sheet on SL pg. 34. ☐ Add information learned from the demonstration on SL pg. 34.	☐ (*Optional*) Complete the copywork or dictation assignment and add it to the Astronomy Notes sheet on SL pg. 34. ☐ (*Optional*) Fill out the record sheet on SL pg. 36 for one of the projects. ☐ (*Optional*) Take Astronomy Quiz #3.
Do	☐ (*Optional*) Make the Uranus model.	☐ (*Optional*) Make the Neptune model.	☐ Do the demonstration entitled "Planetary Orbit."	☐ Work on the Solar System Model.

Chapter 7: List schedule

Supplies Needed	
Demo	• Marble, Smooth pie plate or cake pan
Projects	• Materials will vary based on the type of planet model you choose to make.

Chapter Summary

The chapter opens with the two Triple S agents, Evan DeBlose and Q-Tip, finally reaching the targeted satellite. It then quickly switches to Adrianna Archer, the turncoat, who is now sitting outside of Uncle Cecil's house, watching his every move. Back in space, Blaine and Tracey decide to continue the search for Brett and the Dark Cape by heading out to the next planet, Uranus. After learning about the planet and realizing the duo they seek is not there, they head onto Neptune. Meanwhile, on the I.S.S., the struggle for justice continues and Partha is one step closer to being kicked off the space station. Back at Neptune, the twins learn about the planet, but their focus remains on finding Brett. When they don't see him, they decide to return to the I.S.S., defeated. But close by the Man With No Eyebrows in the Dark Cape has a change of heart and debates about where to take Brett so that he will be seen and rescued. Back at the I.S.S., the twins share their failure with Yang Bo, but as they are speaking, Brett is spotted just outside the receiving room. They bring him onto the space station, and we switch to the Man With No Eyebrows where we find out that he has decided to now be a force for good. The chapter wraps up on the I.S.S. where Brett wakes up and everyone is waiting with bated breath to hear his side of the story.

Essential To-Do's

Read
- ☐ Read the section entitled "Unsurmountable Uranus" of Chapter 7 in *SSA Volume 6: Astronomy*.
- ☐ Read the section entitled "Nearing Neptune" of Chapter 7 in *SSA Volume 6: Astronomy*.

Write
- ☐ Fill out a Astronomy Record Sheet on SL pg. 31 on Uranus.
- ☐ Go over the vocabulary word and enter it into the Astronomy Glossary on SL pg. 94.
- ☐ Fill out a Astronomy Record Sheet on SL pg. 32 on Neptune.
- ☐ Add information learned from the demonstration on SL pg. 34.

Do
- ☐ Do the demo entitled "Planetary Orbit."
- ☐ Work on the Solar System Model.

Optional Extras

Read
- ☐ Read one or all of the assigned pages from the encyclopedia of your choice.
- ☐ Read one of the additional books from your library.

Write
- ☐ Add observations to the Night Sky Journal Sheet on SL pg. 34.
- ☐ Write a narration on the Astronomy Notes Sheet on SL pg. 34.
- ☐ Complete the copywork or dictation assignment and add it to the Astronomy Notes sheet on SL pg. 34.
- ☐ Fill out the record sheet on SL pg. 36 for one of the projects.
- ☐ Write a mission report.   ☐ Take Astronomy Quiz #3.

Do
- ☐ Make the Uranus model.
- ☐ Make the Neptune model.

Chapter 7: Where in the Solar System is Mr. Frye?

Science-Oriented Books

Living Book Spine
- Chapter 7 of *The Sassafras Science Adventures Volume 6: Astronomy* (NOTE—Pluto is mentioned as a dwarf planet in this chapter. However, it is not covered in detail, yet. Dwarf planets, including Pluto, will be covered in Chapter 14.)

Optional Encyclopedia Readings
- *Basher Science Astronomy* pg. 42 (Uranus), pg. 44 (Neptune)
- *Usborne Children's Encyclopedia* pp. 270-271 (Uranus and Neptune)
- *DK First Space Encyclopedia* pp. 72-73 (Distant Twins)
- *Kingfisher Science Encyclopedia* pg. 408 (Uranus), pg. 409 (Neptune)

Additional Living Books
- *Uranus (True Books)* by Christine Taylor-Butler
- *Uranus* by Seymour Simon
- *Uranus: The Ice Planet (Our Solar System)* by Greg Roza
- *The Sideways Planet: Uranus (Amazing Science: Planets)* by Nancy Loewen and Jeff Yesh
- *Neptune* by Seymour Simon
- *Neptune: The Stormy Planet (Our Solar System)* by Greg Roza
- *Planet Neptune (True Books)* by Ann O. Squire
- *Neptune (Scholastic News Nonfiction Readers: Space Science)* by Melanie Chrismer
- *Farthest from the Sun: The Planet Neptune (Amazing Science: Planets)* by Nancy Loewen and Jeff Yesh

Notebooking (SCIDAT Logbook Information)

This week, you can have the students complete a night sky journal sheet. You can also have them fill out the logbook sheets for Uranus and Neptune. Here is the information they could include:

Night Sky Journal

This week, you can look for Uranus and Neptune when you do your night sky observations. The following article has information about how to spot the planets in the night sky:
- https://www.space.com/39240-when-to-see-planets-in-the-sky.html (NOTE—*This article link contains information for spotting the planets in 2019. It will still help with spotting the planets at any time, but you can check the resource page in the introduction of this guide for an updated link.*)

Astronomy Record Sheets
Uranus
Information Learned

- *Uranus is the seventh planet from the sun.*
- *Uranus is a gas giant. It consists mostly of hydrogen and helium, but astronomers believe that it has a solid core.*
- *The planet's blue color is due to the presence of methane gas.*

- *Uranus has a system of eleven rings, which are all quite thin.*
- *It also has over twenty confirmed moons, the largest of which is half the size of earth's moon.*
- *Uranus is unique in our solar system because it orbits the sun on its side.*
- *A year on Uranus takes about 84 Earth years.*
- *A day takes a little over 17 Earth hours.*

Neptune
Information Learned

- *Neptune is the farthest planet from the sun and is about four times bigger than Earth.*
- *It appears only as a tiny blue star when we look at it through a telescope from Earth.*
- *Neptune is another gas giant with a small rocky core.*
- *It is composed mainly of hydrogen and helium, but it also has methane gas, which gives the planet its blue color, and ammonia gas, which forms the pure white clouds on the surface.*
- *Neptune is the planet that is known to have the worst storms in our solar system.*
- *In 1989, the probe Voyager 2 sent back images of Neptune that depicted a raging storm which was named the Great Dark Spot. But then in 1994, the spot mysteriously disappeared.*
- *Neptune has more than ten confirmed moons, the largest of which is Triton.*
- *A year on Neptune takes 165 Earth years.*
- *A day takes only a little over 16 Earth hours.*

Vocabulary

Have the older students look up the following term in the glossary in the Appendix on pp. 137-138 or in a science encyclopedia. Then, have them copy each definition onto a blank index card or into their SCIDAT logbook.

- **Gas Giant** – A large planet in our solar system that is composed mainly of gas.

Scientific Demonstration: Planetary Orbit

Materials
- ☑ Marble
- ☑ Smooth pie plate or cake pan

Procedure
1. Set the pie plate or cake pan on a smooth surface.
2. Have the students set the marble on the edge of the pan and push it so that it travels around the edge of the pan. Have them observe what the marble does.
3. Have the students touch the marble and then observe what happens to the marble's path.

Explanation

The students should see that the marble continues to travel in a circular path around the edge of the pie plate even when they are not pushing it. When they touch the marble, it should slow down or stop. This is because an object in motion will stay in motion unless a force acts upon it. This is known as inertia, or the resistance that an object has to any change in its motion. The same principle is true in our solar system. The planets continue to orbit, or move around the sun, because space is a vacuum and there is not a force great enough to stop their motion. Inertia keeps the planets orbiting the sun, while gravity from the sun keeps the path of the planet in the same relative position.

Take it Further
Have the students see how both gravity and inertia affects the planet's orbit. You will need at least two people and a jump rope. Have each person hold an end of the jump rope so that the rope is taut. Have one person be the stationary sun person who stands relatively still, just turning round in a circle to follow the runner without wrapping up the rope. Have the other person be the planet runner who tries to run in a straight line, while the stationary sun person makes sure that the rope stays taut. (*You should see the that the planet runner ends up going in a circle around the stationary sun person. This is how both gravity and inertia work together to keep the planets in orbit.*)

Multi-Week Projects and Activities

Multi-week Projects
- **Solar System Model** – This week, the students will add the last two planets, Uranus and Neptune, to their solar system model. Have the students cut our the planets from copy paper or construction paper using the diameters below. Have them color the planets using a picture from the suggested books or from the Internet. When they are done, have them add the planets to their solar system using the distances below.

  Uranus
  - Diameter (wall version): 8 in
  - Distance (wall version): 6 feet, 5 in
  - Diameter (lap version): 4 cm
  - Distance (lap version): 38.6 cm

  Neptune
  - Diameter (wall version): 7.5 in
  - Distance (wall version): 10 feet, 1 in
  - Diameter (lap version): 3.8 cm
  - Distance (lap version): 60.6 cm

Activities for This Week
- **Uranus Model** – Have the students make a 3D model of Uranus. You can have them paint a Styrofoam ball, or you can have them make a paper-mâché model. (**NOTE**—*See the directions found on Appendix pg. 129 for the paper-mâché plane*t.)
- **Neptune Model** – Have the students make a 3D model of Neptune. You can have them paint a Styrofoam ball, or you can have them make a paper-mâché model. (**NOTE**—*See the directions found on Appendix pg. 129 for the paper-mâché plane*t.)
- **Mission Report** – Have the students research and write a brief report about the *Voyager 2* mission to Uranus and Neptune. This report should include the dates of the mission, several pictures, and a summary of what was found.

Memorization

Copywork/Dictation
- **Copywork Sentence**
  Uranus orbits the sun on its side. Neptune has the worst storms in our solar system.
- **Dictation Selection**
  Uranus and Neptune are both gas giants. Uranus is composed mainly of hydrogen and helium with a solid core. The seventh planet has a blue color due to the presence of methane gas. Neptune is composed mainly of hydrogen and helium with a small rocky core. The eighth planet also has methane gas, which gives it a blue color, and ammonia gas, which forms the pure white clouds on the surface.

Quiz Information
This week, you can give the students a quiz based on what they learned in chapters 6 and 7. You can find this quiz in the Appendix on pg. 145.

Quiz #3 Answers
1. Jupiter, Saturn, Uranus, Neptune
2. Largest
3. False (*Jupiter's red spot is constantly moving.*)
4. Rocks, dust, and ice
5. Light
6. False (*Uranus has over twenty confirmed moons.*)
7. True
8. Neptune

Chapter 7 Notes

Chapter 8: Grid Schedule

	Supplies Needed
Demo	• Glass bowl, Cooking oil, Piece of paper with words on it, Magnifying glass
Projects	• White tissue paper or white chalk pastel • White chalk pastel or crayon, Black construction paper, Silver glitter

Chapter Summary

The chapter opens with the twins and the rest of the I.S.S. crew solving the mystery behind Brett's disappearance. Blaine and Tracey leave the I.S.S. and return to *Ulysses-1*, which takes them back to Earth where they head off to their next location in Hawaii. We also learn that the Man With No Eyebrows is returning to his Siberian lab. The twins arrive on the beach where a crowd has formed around a man building a sand castle. The man turns out to be their local expert, J.P. Jungos, who is obviously a burn victim. J.P. is not only building a sand castle, but he is also adding a poem about galaxies to his "Darling." The twins learn about galaxies before heading down the beach to get a meal and find a spot to sleep for the night. After an evening of rest, they learn about telescopes from J.P. Jungos and decide to join him on his journey of a lifetime to the Mauna Kea Observatories. The chapter ends with J.P. sharing a bit about his journey and his "Darling."

Weekly Schedule

	Day 1	Day 2	Day 3	Day 4
Read	☐ Read the section entitled "Gawking at Galaxies" of Chapter 8 in *SSA Volume 6: Astronomy*.	☐ Read the section entitled "Towering Telescopes" of Chapter 8 in *SSA Volume 6: Astronomy*.	☐ *(Optional)* Read one or all of the assigned pages from the encyclopedia of your choice.	☐ *(Optional)* Read one of the additional books from your library.
Write	☐ Fill out a Astronomy Record Sheet on SL pg. 39 on galaxies. ☐ Go over the vocabulary words and enter them into the Astronomy Glossary on SL pg. 94.	☐ Fill out a Astronomy Record Sheet on SL pg. 40 on telescopes. ☐ *(Optional)* Add observations to the Night Sky Journal Sheet on SL pg. 37.	☐ *(Optional)* Write narration on the Astronomy Notes Sheet on SL pg. 43. ☐ Add information learned from the demonstration on SL pg. 43.	☐ *(Optional)* Complete the copywork or dictation assignment and add it to the Astronomy Notes sheet on SL pg. 43. ☐ *(Optional)* Fill out the record sheet on SL pg. 45 for one of the projects.
Do	☐ *(Optional)* Make the Milky Way art.	☐ *(Optional)* Do the Hubble Telescope project.	☐ Do the demonstration entitled "Magnify."	☐ Work on the Solar System Model.

Chapter 8: List schedule

	Supplies Needed
Demo	• Glass bowl, Cooking oil, Piece of paper with words on it, Magnifying glass
Projects	• White tissue paper or white chalk pastel • White chalk pastel or crayon, Black construction paper, Silver glitter

Chapter Summary

The chapter opens with the twins and the rest of the I.S.S. crew solving the mystery behind Brett's disappearance. Blaine and Tracey leave the I.S.S. and return to *Ulysses-1*, which takes them back to Earth where they head off to their next location in Hawaii. We also learn that the Man With No Eyebrows is returning to his Siberian lab. The twins arrive on the beach where a crowd has formed around a man building a sand castle. The man turns out to be their local expert, J.P. Jungos, who is obviously a burn victim. J.P. is not only building a sand castle, but he is also adding a poem about galaxies to his "Darling." The twins learn about galaxies before heading down the beach to get a meal and find a spot to sleep for the night. After an evening of rest, they learn about telescopes from J.P. Jungos and decide to join him on his journey of a lifetime to the Mauna Kea Observatories. The chapter ends with J.P. sharing a bit about his journey and his "Darling."

Essential To-Do's

Read
☐ Read the section entitled "Gawking at Galaxies" of Chapter 8 in *SSA Volume 6: Astronomy*.
☐ Read the section entitled "Towering Telescopes" of Chapter 8 in *SSA Volume 6: Astronomy*.

Write
☐ Fill out a Astronomy Record Sheet on SL pg. 39 on galaxies.
☐ Go over the vocabulary words and enter them into the Astronomy Glossary on SL pg. 94.
☐ Fill out a Astronomy Record Sheet on SL pg. 40 on telescopes.
☐ Add information learned from the demonstration on SL pg. 43.

Do
☐ Do the demonstration entitled "Magnify."
☐ Work on the Solar System Model.

Optional Extras

Read
☐ Read one or all of the assigned pages from the encyclopedia of your choice.
☐ Read one of the additional books from your library.

Write
☐ Add observations to the Night Sky Journal Sheet on SL pg. 43.
☐ Write a narration on the Astronomy Notes Sheet on SL pg. 43.
☐ Complete the copywork or dictation assignment and add it to the Astronomy Notes sheet on SL pg. 43.
☐ Fill out the record sheet on SL pg. 45 for one of the projects.

Do
☐ Make the Milky Way art.
☐ Do the Hubble Telescope project.

Chapter 8: Back To Earth

Science-Oriented Books

Living Book Spine
- Chapter 8 of *The Sassafras Science Adventures Volume 6: Astronomy*

Optional Encyclopedia Readings
- *Basher Science Astronomy* pg. 90 (The Milky Way), pg. 108 (Hubble Space Telescope)
- *Usborne Children's Encyclopedia* pg. 276-277 (Galaxies)
- *DK First Space Encyclopedia* pp. 16-19 (Great Galaxies and The Milky Way), pp. 10-13 (Observatories and Radio Telescopes)
- *Kingfisher Science Encyclopedia* pg 390-391 (Galaxies), pg. 270-271 (Telescopes)

Additional Living Books
- *The Milky Way (Exploring Space)* by Martha E. H. Rustad and Ilia I. Roussev
- *The Milky Way (Galaxy)* by Gregory L. Vogt
- *Looking Through a Telescope (Rookie Read-About Science)* by Linda Bullock
- *Telescopes (First Facts: Science Tools)* by Adele Richardson
- *The Hubble Telescope (Blastoff! Readers: Exploring Space)* by Derek Zobel

Notebooking (SCIDAT Logbook Information)

This week, you can have the students complete a night sky journal sheet. You can also have them fill out the logbook sheets for galaxies and telescopes. Here is the information they could include:

Night Sky Journal
This week, you can use a telescope for evidence of galaxies when you do your night sky observations. The following article has information about how to spot the planets in the night sky:
⇨ *On clear nights, you can see bands and trails of light and dark up in the sky. These are actually clusters of stars and trails of dust within the bands of the spiral galaxy.*

Astronomy Record Sheets
Galaxies
Information Learned
⇨ *A galaxy is a cluster of millions of stars, gas, and dust grouped together in a shape that is held together by gravity.*
⇨ *There are four main shapes of galaxies: spiral, elliptical, barred spiral, and irregular.*
  1. *Spiral galaxies have a pinwheel shape with a bulge and thin disk in the center. They have lots of gas and dust with both young and old stars.*
  2. *Elliptical galaxies have a round or oval shape with a bulge in the center, but no disk. They have a little cool dust with both young and old stars.*
  3. *Barred spiral galaxies have a pinwheel shape with a bar of gas, dust, and stars running through the center. They usually have lots of gas and dust with both young and old stars.*

4. Irregular galaxies contain a hodgepodge of shapes, basically anything that is not spiral or elliptical, meaning they have no regular shape. They normally have lots of gas and dust with both young and old stars.

⇨ The galaxy within which the Earth is found is called the Milky Way. It is part of the Local Group cluster, which contains about 30 galaxies. Astronomers believe the Milky Way is spiral shaped.

Telescope

Information Learned

⇨ Telescopes are designed to make distant objects look closer.
⇨ There are two main types of optical (light-collecting) telescopes:
   1. Refractor telescopes – These telescopes use a curved mirror to collect light and sent it to another mirror, which focuses it in front of the eyepiece that magnifies it.
   2. Reflector telescopes – These telescopes use lens to collect light towards the eyepiece, which magnifies the image.
⇨ In an optical telescope, the more light that hits the eyepiece, the brighter and clearer the image will be.
⇨ We have telescopes on Earth in observatories and some in space, like the Hubble Telescope.
⇨ Telescopes out in space don't have to deal with the haze and pollution of our atmosphere, so the images sent back are much clearer.
⇨ In addition to light, telescopes can also collect radio and energy waves. Radio telescopes collect radio waves is a dish and reflect them to a receiver, where they are processed by a computer and turned into images.

Vocabulary

Have the older students look up the following term in the glossary in the Appendix on pp. 137-138 or in a science encyclopedia. Then, have them copy each definition onto a blank index card or into their SCIDAT logbook.

◌ **Refraction** – The bending of light as it passes through a different substance.
◌ **Telescope** – An instrument to look at things far away in space.

Scientific Demonstration: Magnify

Materials
☑ Glass bowl
☑ Cooking oil (vegetable or coconut work best, as they are more clear)
☑ Piece of paper with words on it
☑ Magnifying glass

Procedure
1. Place the piece of paper with the words on it on a desk or table. Have the students place the glass bowl on top of the paper and observe how the words appear.
2. Next, have the students fill the glass bowl halfway with the vegetable oil and observe how the words change.
3. Finally, have the students remove the glass and use the magnifying glass to observe what the words look like.
4. Once they are done, ask the students the following:
   ? What happened to the words on the paper when you place the bowl on top?
   ? What happened when you added oil to the bowl?
   ? What happened when you used the magnifying glass to look at the words?

Explanation

The students should see that the words on the page are magnified when the bowl is full of water and when they look at the paper with the magnifying glass. When you filled the bowl with oil the words on the page got bigger. The same was true when you used the magnifying glass. Light that bounced off our page was bent in the oil and through the lens in the magnifying glass, which made the words bigger. A telescope works in the same way. There are lenses inside that refract, or bend, the light so that objects appear bigger and we can see more detail.

Take it Further

Have the students do their own night sky spotting using a telescope or binoculars. Here are a few tips to help you guide their spotting adventure:

- https://elementalscience.com/blogs/podcast/episode-9

Multi-Week Projects and Activities

Multi-week Projects

- **Solar System Model** – This week, you can have the students add a bit of milky haze to represent the Milky Way galaxy. You can use tissue paper for the wall model and white chalk pastel for the lap version.

Activities For This Week

- **Milky Way Art** – Have the students make their own Milky Way drawing using a white chalk pastel or crayon on black construction paper. Then have them use glue to trace the lines and sprinkle silver glitter over them.
- **Hubble Telescope** – Have the students learn more about the Hubble Telescope. Have them learn more about what the telescope does and view some of the pictures it has taken. You can do this through the following website:
    - http://hubblesite.org/the_telescope/

If they are older, have them write a brief paragraph about what they have learned.

Memorization

Copywork/Dictation

☞ **Copywork Sentence**

A galaxy is a cluster of millions of stars, gas, and dust. A galaxy is held together by gravity.

☞ **Dictation Selection**

A galaxy is a cluster of millions of stars, gas, and dust grouped together in a shape that is held together by gravity. There are four main shapes of galaxies: spiral, elliptical, barred spiral, and irregular. The Earth is found in the Milky Way galaxy, which astronomers believe is a spiral-shaped galaxy.

Chapter 8 Notes

Chapter 9: Grid Schedule

	Supplies Needed
Demo	• Small mirror, Small flashlight, A dark room
Projects	• Bottle caps, Toothpicks, Thin cardboard, A small juice box, Glittered blue decorative card stock, Gold and silver paint, 12x12-inch Foam piece, Aluminum foil, Glue, 1/4-inch Wooden dowel, Scissors, Pencil, Measuring tape, Clear tape • Smartphone

Chapter Summary

The chapter opens with the Man With No Eyebrows arriving back at his Siberian lab, but Adrianna is not there to greet him. We head back to Blaine and Tracey, who are heading out to hike through the Hawaiian rainforest with their local expert, J.P. Jungos. The twins learn about the fire that led to J.P.'s burns and the loss of his wife. After a full day of hiking, they stop for a rest and the twins awaken the next morning to J.P. talking about satellites. The three resume their hike to the Mauna Kea Observatories. Meanwhile, we learn that Agent DeBlose and Q-tip were able to repair the Triple S satellite. We also see all the newfound goodwill disappear as the Man With No Eyebrows finds out that Adrianna Archer has left for good. Back in Hawaii, the twins and their local expert have a run-in with a crank tour guide, Peter Karko. They are rescued by Dr. Ellison Ocampo, who gives them a proper tour of the observatory. The chapter ends with the twins learning about space probes as J.P.'s journey comes to an end.

Weekly Schedule

	Day 1	Day 2	Day 3	Day 4
Read	☐ Read the section entitled "Satellites Switchbacks" of Chapter 9 in *SSA Volume 6: Astronomy*.	☐ Read the section entitled "Startling Space Probes" of Chapter 9 in *SSA Volume 6: Astronomy*.	☐ (*Optional*) Read one or all of the assigned pages from the encyclopedia of your choice.	☐ (*Optional*) Read one of the additional books from your library.
Write	☐ Fill out a Astronomy Record Sheet on SL pg. 41 on satellites. ☐ Go over the vocabulary words and enter them into the Astronomy Glossary on SL pp. 94-95.	☐ Fill out a Astronomy Record Sheet on SL pg. 42 on space probes. ☐ (*Optional*) Add observations to the Night Sky Journal Sheet on SL pg. 44.	☐ (*Optional*) Write narration on the Astronomy Notes Sheet on SL pg. 44. ☐ Add information learned from the demonstration on SL pg. 44.	☐ (*Optional*) Complete the copywork or dictation assignment and add it to the Astronomy Notes sheet on SL pg. 44. ☐ (*Optional*) Fill out the record sheet on SL pg. 46 for one of the projects. ☐ (*Optional*) Take Astronomy Quiz #4.
Do	☐ (*Optional*) Make the Satellite model.	☐ (*Optional*) Go geocaching.	☐ Do the demonstration entitled "Reflection Direction."	☐ Work on the Solar System Model.

Chapter 9: List Schedule

	Supplies Needed
Demo	• Small mirror, Small flashlight, A dark room
Projects	• Bottle caps, Toothpicks, Thin cardboard, A small juice box, Glittered blue decorative card stock, Gold and silver paint, 12x12-inch Foam piece, Aluminum foil, Glue, 1/4-inch Wooden dowel, Scissors, Pencil, Measuring tape, Clear tape • Smartphone

Chapter Summary

The chapter opens with the Man With No Eyebrows arriving back at his Siberian lab, but Adrianna is not there to greet him. We head back to Blaine and Tracey, who are heading out to hike through the Hawaiian rainforest with their local expert, J.P. Jungos. The twins learn about the fire that led to J.P.'s burns and the loss of his wife. After a full day of hiking, they stop for a rest and the twins awaken the next morning to J.P. talking about satellites. The three resume their hike to the Mauna Kea Observatories. Meanwhile, we learn that Agent DeBlose and Q-tip were able to repair the Triple S satellite. We also see all the newfound goodwill disappear as the Man With No Eyebrows finds out that Adrianna Archer has left for good. Back in Hawaii, the twins and their local expert have a run-in with a crank tour guide, Peter Karko. They are rescued by Dr. Ellison Ocampo, who gives them a proper tour of the observatory. The chapter ends with the twins learning about space probes as J.P.'s journey comes to an end.

Essential To-Do's

Read
- ☐ Read the section entitled "Satellites Switchbacks" of Chapter 9 in *SSA Volume 6: Astronomy*.
- ☐ Read the section entitled "Startling Space Probes" of Chapter 9 in *SSA Volume 6: Astronomy*.

Write
- ☐ Fill out a Astronomy Record Sheet on SL pg. 41 on satellites.
- ☐ Go over the vocabulary words and enter them into the Astronomy Glossary on SL pp. 94-95.
- ☐ Fill out a Astronomy Record Sheet on SL pg. 42 on space probes.
- ☐ Add information learned from the demonstration on SL pg. 44.

Do
- ☐ Do the demo entitled "Reflection Direction."
- ☐ Work on the Solar System Model.

Optional Extras

Read
- ☐ Read one or all of the assigned pages from the encyclopedia of your choice.
- ☐ Read one of the additional books from your library.

Write
- ☐ Add observations to the Night Sky Journal Sheet on SL pg. 44.
- ☐ Write a narration on the Astronomy Notes Sheet on SL pg. 44.
- ☐ Complete the copywork or dictation assignment and add it to the Astronomy Notes sheet on SL pg. 44.
- ☐ Fill out the record sheet on SL pg. 46 for one of the projects.
- ☐ Take Astronomy Quiz #4.

Do
- ☐ Make the satellite model.
- ☐ Go geocaching.

Chapter 9: Hawaiian Dreams and Sightings

Science-Oriented Books

Living Book Spine
- Chapter 9 of *The Sassafras Science Adventures Volume 6: Astronomy*

Optional Encyclopedia Readings
- *Basher Science Astronomy* pg. 19 (Space Junk), pg. 51 (Voyagers)
- *Usborne Children's Encyclopedia* pp. 254-255 (Satellites and Space Probes)
- *DK First Space Encyclopedia* pp. 24-25 (Exploring Space), pp. 44-45 (Artificial Satellites), pp. 86-87 (Space Debris)
- *Kingfisher Science Encyclopedia* pp. 418-419 (Exploring Space), pp. 424-425 (Artificial Satellites)

Additional Living Books
- *Satellites and Space Probes (Eye on the Universe)* by Niki Walker
- *All About Satellites (Blast Off!)* by Miriam Gross
- *Satellites (Let's See Library: Communication)* by Darlene R. Stille

Notebooking (SCIDAT Logbook Information)

This week, you can have the students complete a night sky journal sheet. You can also have them fill out the logbook sheets for satellites and space probes. Here is the information they could include:

Night Sky Journal
This week, you can look for satellites when you do your night sky observations. The following article has information about how to spot the satellites in the night sky:
- https://wxguys.ssec.wisc.edu/2013/07/02/can-we-see-satellites-at-night/

Astronomy Record Sheets
Satellites
Information Learned

- The word "satellite" comes from the Latin word for attendant.
- A satellite is an object, either man-made or natural, that orbits something bigger than itself.
- The moon is technically a natural satellite of the Earth.
- Some satellites look out into space as they orbit around the Earth; others look down on Earth as they orbit the Earth.
- There are thousands of man-made satellites orbiting the Earth; some are working, some are not. Those that are not working are called space litter.
- There are several different types of man-made satellites.
    1. Communication satellites – These satellites help capture and transfer radio waves.
    2. Navigation satellites – These satellites allow use to establish position on the earth, i.e. GPS.
    3. Military satellites – These satellites are used by the military for communication, navigation, and other tasks.

4. *Resource satellites* – *These satellites take pictures of the earth's natural resources for scientists to turn into maps.*
5. *Weather satellites* – *These satellites help scientists to learn about and forecast weather.*

Space Probes
Information Learned

⇨ *Space probes are similar to man-made satellites, except they travel around space and don't just orbit the Earth.*
⇨ *These unmanned spacecrafts typically carry cameras and send back images of the distant places they visit.*
⇨ *Space probes have visited all the planets in our solar system. Some have even landed on the planets.*
⇨ *The space probes Voyager 1 and Voyager 2 have gone further than we have ever gone before.*
⇨ *Space probes have helped scientists learn more about asteroids, black holes, comets, the planets in our solar system, and the sun.*
⇨ *Sputnik 1 was the very first space probe to go out into space. It was launched in 1957 by the Soviet Union.*
⇨ *The United States launched its first space probe, Explorer 1, a year later.*

Vocabulary

Have the older students look up the following term in the glossary in the Appendix on pp. 137-138 or in a science encyclopedia. Then, have them copy each definition onto a blank index card or into their SCIDAT logbook.

- **Reflection** – The change in direction of light rays that occurs when it hits an object and bounces off.
- **Satellite** – An object, either man-made or natural, that orbits something bigger than itself.

Scientific Demonstration: Reflection Direction

Materials
- ☑ Small mirror
- ☑ Small flashlight
- ☑ A dark room

Procedure
1. In a dark room, have a student hold the mirror as you face him or her and shine the flashlight directly at the mirror. Have the student observe how the beam of light bounces off the mirror and appears on the wall behind you.
2. Now, have the students tilt the mirror up or down.
   ❓ What happened to the beam of light on the wall?
3. Then, have the students tilt the mirror left or right.
   ❓ What happened to the beam of light on the wall this time?

Explanation
The students should see that the beam of light moved when the mirror was tilted. The direction of the reflected beam is dependent and where it appears on the wall on the angle of the mirror. Satellites work in much the same way. Radio waves are sent from Earth to the satellite. The satellite then relay, or send back, the waves to Earth in another location.

Take it Further
Have the students repeat the demonstration with the flashlight further away from the mirror in a larger room. How did the distance affect the direction of the beam of light? (*The students should see the further away the flashlight is from the mirror, the greater the angle of reflection of the beam of light.*)

Multi-Week Projects and Activities

Multi-week Projects
✂ **Solar System Model** – This week, have the students add a satellite and a space probe to their wall model of the solar system. You can have them draw their own or use a copy of the images on their logbook pages. Have them place the satellite image as if it is orbiting Earth. Have them place the space probe image near one of the outer planets. (**NOTE**—*If you are creating a lap-sized version of the solar system, skip this assignment as the satellites and space probes would be almost microscopic.*)

Activities for This Week
✂ **Satellite Model** – Have the students make a model satellite using bottle caps, toothpicks, thin cardboard, a small juice box, glittered blue decorative card stock, gold and silver paint, 12x12-inch foam piece, aluminum foil, glue, 1/4-inch wooden dowel, scissors, pencil, measuring tape, and clear tape. You can find the directions for this project here:
🖱 https://www.ehow.com/how_5006567_make-model-satellite.html

(**NOTE**—*If you don't have access to all these supplies, you can have your students use LEGO bricks and a picture of a satellite to build their own creations.*)

✂ **Geocaching** – Have the students put all those global positioning satellites (GPS) to work with a smartphone as you go geocaching! Directions for geocaching can be found here:
🖱 https://www.geocaching.com/play

Memorization

Copywork/Dictation
☞ **Copywork Sentence**
Man-made satellites orbit the Earth. Space probes travel around space.

☞ **Dictation Selection**
A satellite is an object, either man-made or natural, that orbits something bigger than itself. The moon is technically a natural satellite of the Earth. Some satellites look out into space as they orbit around the Earth, while others look down on Earth as they orbit the Earth. Space probes are similar to man-made satellites, except they travel around space and don't just orbit the Earth.

Quiz Information
This week, you can give the students a quiz based on what they learned in chapters 8 and 9. You can find this quiz in the Appendix on pg. 147.

Quiz #4 Answers
1. C, A, D, B
2. Milky Way
3. Bigger
4. Bends
5. False (*Space probes visit other planets, while satellites orbit the earth.*)
6. All of the answers should be circled.
7. Unmanned
8. Bounces

Chapter 9 Notes

Chapter 10: Grid Schedule

	Supplies Needed
Demo	• Large marshmallows, Chocolate squares, Graham crackers, Foil, Cardboard box, Plastic wrap
Projects	• White glue, Food coloring, Toothpicks, Yogurt container lid, Hole punch, String • Colored pencils or magazine pictures

Chapter Summary

The chapter opens with the twins honoring the memory of J.P. Jungos as they get ready to zip to their next location – Washington, DC – where their local expert will be Paul Sims. For a second after their arrival, the twins fear that they have landed back in space. They quickly realize that they are in a museum model and as they step out, they learn about the sun as Paul shares with a museum visitor. Moments later REESE and Summer appear: it turns out that Paul Sims is an old classmate of Summer and Cecil. Summer delivers a piece of an asteroid to Paul for the museum and the twins learn about the day/night cycle. Paul is getting ready to show them a movie when they are interrupted by Shine-O-Mite, the custodial crew, and the security guards, Wiggles and Fidget. After that, Summer and Paul meet with the museum bigwigs about the space rock and Blaine and Tracey are left to wander the museum. They are approached by the custodial crew, who say they are in danger. But when the twins refuse to go with the crew, the crew kidnaps them. We then flash to the Triple S agents using their repaired satellite to locate the turncoat, Adrianna Archer. The chapter ends with the Man With No Eyebrows and the Triple S simultaneously finding out that Adrianna is walking into the Left-Handed Turtle Market.

Weekly Schedule

	Day 1	Day 2	Day 3	Day 4
Read	☐ Read the section entitled "Solar Gumballs" of Chapter 10 in *SSA Volume 6: Astronomy*.	☐ Read the section entitled "Daytime Here, Nighttime There" of Chapter 10 in *SSA Volume 6: Astronomy*.	☐ *(Optional)* Read one or all of the assigned pages from the encyclopedia of your choice.	☐ *(Optional)* Read one of the additional books from your library.
Write	☐ Fill out a Astronomy Record Sheet on SL pg. 49 on the sun. ☐ Go over the vocabulary word and enter it into the Astronomy Glossary on SL pg. 95.	☐ Fill out a Astronomy Record Sheet on SL pg. 50 on the day/night cycle. ☐ *(Optional)* Add observations to the Night Sky Journal Sheet on SL pg. 47.	☐ *(Optional)* Write narration on the Astronomy Notes Sheet on SL pg. 53. ☐ Add information learned from the demonstration on SL pg. 53.	☐ *(Optional)* Complete the copywork or dictation assignment and add it to the Astronomy Notes sheet on SL pg. 53. ☐ *(Optional)* Fill out the record sheet on SL pg. 55 for one of the projects.
Do	☐ *(Optional)* Make suncatchers.	☐ *(Optional)* Do the Day/Night collage project.	☐ Do the demonstration entitled "Solar S'mores."	☐ Work on the Solar System Model.

Chapter 10: List Schedule

	Supplies Needed
Demo	• Large marshmallows, Chocolate squares, Graham crackers, Foil, Cardboard box, Plastic wrap
Projects	• White glue, Food coloring, Toothpicks, Yogurt container lid, Hole punch, String • Colored pencils or magazine pictures

Chapter Summary

The chapter opens with the twins honoring the memory of J.P. Jungos as they get ready to zip to their next location – Washington, DC – where their local expert will be Paul Sims. For a second after their arrival, the twins fear that they have landed back in space. They quickly realize that they are in a museum model and as they step out, they learn about the sun as Paul shares with a museum visitor. Moments later REESE and Summer appear: it turns out that Paul Sims is an old classmate of Summer and Cecil. Summer delivers a piece of an asteroid to Paul for the museum and the twins learn about the day/night cycle. Paul is getting ready to show them a movie when they are interrupted by Shine-O-Mite, the custodial crew, and the security guards, Wiggles and Fidget. After that, Summer and Paul meet with the museum bigwigs about the space rock and Blaine and Tracey are left to wander the museum. They are approached by the custodial crew, who say they are in danger. But when the twins refuse to go with the crew, the crew kidnaps them. We then flash to the Triple S agents using their repaired satellite to locate the turncoat, Adrianna Archer. The chapter ends with the Man With No Eyebrows and the Triple S simultaneously finding out that Adrianna is walking into the Left-Handed Turtle Market.

Essential To-Do's

Read
- ☐ Read the section entitled "Solar Gumballs" of Chapter 10 in *SSA Volume 6: Astronomy*.
- ☐ Read the section entitled "Daytime Here, Nighttime There" of Chapter 10 in *SSA Volume 6: Astronomy*.

Write
- ☐ Fill out a Astronomy Record Sheet on SL pg. 49 on the sun.
- ☐ Go over the vocabulary word and enter it into the Astronomy Glossary on SL pg. 95.
- ☐ Fill out a Astronomy Record Sheet on SL pg. 50 on the day/night cycle.
- ☐ Add information learned from the demonstration on SL pg. 53.

Do
- ☐ Do the demonstration entitled "Solar S'mores."
- ☐ Work on the Solar System Model.

Optional Extras

Read
- ☐ Read one or all of the assigned pages from the encyclopedia of your choice.
- ☐ Read one of the additional books from your library.

Write
- ☐ Add observations to the Night Sky Journal Sheet on SL pg. 47.
- ☐ Write a narration on the Astronomy Notes Sheet on SL pg. 53.
- ☐ Complete the copywork or dictation assignment and add it to the Astronomy Notes sheet on SL pg. 53.
- ☐ Fill out the record sheet on SL pg. 55 for one of the projects.

Do
- ☐ Make suncatchers.
- ☐ Do the Day/Night collage project.

Chapter 10: The National Air and Space Museum

Science-Oriented Books

Living Book Spine
- 📖 Chapter 10 of *The Sassafras Science Adventures Volume 6: Astronomy*

Optional Encyclopedia Readings
- 🔍 *Basher Science Astronomy* pg. 6 (Sun)
- 🔍 *Usborne Children's Encyclopedia* pp. 10-11 (Day and Night), pp. 262-263 (The Sun)
- 🔍 *DK First Space Encyclopedia* pp. 52-53 (The Sun)
- 🔍 *Kingfisher Science Encyclopedia* pp. 394-395 (The Sun)

Additional Living Books
- 📖 *The Sun: Our Nearest Star (Let's-Read-and-Find-Out)* by Franklyn M. Branley and Edward Miller
- 📖 *The Sun* by Seymour Simon
- 📖 *Day and Night (First Step Nonfiction: Discovering Nature's Cycles)* by Robin Nelson
- 📖 *Day and Night (Patterns in Nature)* by Margaret Hall and Jo Miller
- 📖 *What Makes Day and Night (Let's-Read-and-Find-Out Science 2)* by Franklyn M. Branley and Arthur Dorros

Notebooking (SCIDAT Logbook Information)

This week, you can have the students complete a night sky journal sheet. You can also have them fill out the logbook sheets for the sun and the day/night cycle. Here is the information they could include:

Night Sky Journal
This week, you can have your students observe the change from night to day, either during the sunrise or sunset, for your night sky observations.

Astronomy Record Sheets
Sun
Information Learned

- ⇨ The sun is actually a fiery star at the center of our solar system. It is a giant ball of exploding gas.
- ⇨ It is so big that a million little Earths could fit inside of the sun.
- ⇨ The sun is super-hot, over 10,000 degrees Fahrenheit.
- ⇨ It takes the sun's heat about eight minutes to reach Earth and life exists on Earth because we are at just the right distance from the sun.
- ⇨ The surface of the sun is constantly in the motion and because of this there are often darker, or cooler, spots, which we call sunspots.
- ⇨ There are also white, or hotter, spots that we call faculae.
- ⇨ Another feature on the surface of the sun is a prominence, which is a mass of burning gas that sweeps up, almost like a leaping flame of fire.
- ⇨ Solar flares are very similar to prominences, except that travel much farther from the surface of the sun
- ⇨ We have sent out space probes Ulysses, SOHO, and TRACE to study the sun.

Day/Night
Information Learned

- As the Earth turns, or rotates, day changes to night on one side of the globe. On the other side, night changes to day with the rotation of our planet. We call this the day/night cycle.
- The day/night cycle occurs because as the Earth turns, different parts of the surface face the sun.
- In the morning, our side of the Earth is turning to face the sun. This is when we see the sunrise, which ushers in a new day.
- In the evening, our side of the Earth is rotating away from the sun. This is when we see the sunset, which marks the beginning of a new night.
- The day-night cycle takes about twenty-four hours to complete.
- Even on cloudy days, the sun is shining, but you can't feel it due to the clouds. You can also block out the sun's light, which is what creates your shadow.
- The only time it is dark during the day is when the moon blocks out the sun's light, which we call an eclipse.

Vocabulary

Have the older students look up the following term in the glossary in the Appendix on pp. 137-138 or in a science encyclopedia. Then, have them copy each definition onto a blank index card or into their SCIDAT logbook.

- **SOLAR WIND** – A stream of tiny particles that blow off the sun and into space.

Scientific Demonstration: Solar S'mores

Materials
- ☑ Large marshmallows
- ☑ Chocolate squares
- ☑ Graham crackers
- ☑ Foil
- ☑ Cardboard box
- ☑ Plastic wrap

Procedure

1. Have the students line the inside of the cardboard box with foil so that the shiny side is facing up. When done, move the box to a sunny place outdoors. (*If you are doing this demonstration during the colder months, you can place the box in a sunny spot inside; just be aware that most of the windows today are designed to reduce the amount of the sun's energy that enters the home. In other words, the demonstration will take quite a bit longer.*)
2. Then, place the a graham cracker square on the foil. (*You can do more than one, just make sure that they are about a half-inch apart.*) Top the graham cracker square with a marshmallow and place a chocolate square on top of the marshmallow.
3. Cover the shoebox with plastic wrap to keep out any unwanted diners and to help trap the heat from the sun. Check the box every fifteen minutes or so to see if there have been any changes. Each time you check on the box, ask the students:
   - ? Do you see any changes in the chocolate? In the marshmallows? In the graham cracker?
4. When the chocolate melts and the marshmallows are soft, the students can eat and enjoy the solar s'mores!

Explanation
The students should see that the chocolate and the marshmallow get soft and melt, creating a warm, delicious s'more. This is because the energy from the sun's rays were used to heat what was inside the box. In the box, the foil reflected the sun's rays and the plastic wrap helped to trap the heat. That heat was then used to toast the s'mores inside. We can also use the energy of the sun to do things, like power a solar car, heat up water, and tell time.

Take it further
Have the students make a sundial using the directions found on the following website:
  http://www.skyandtelescope.com/observing/make-your-own-sundial/

Multi-Week Projects and Activities

Multi-week Projects
- **SOLAR SYSTEM MODEL** – This week, the students will add features, such as sunspots, faculae, and solar flares, to the sun in the solar system model.

Activities For This Week
- **SUNCATCHERS** – Have the students make suncatchers to enjoy during the day. You will need the following supplies: white glue, food coloring, toothpicks, yogurt container lid, a hole punch, and string. Have the students pour enough white glue to cover the yogurt container lid. Add a few drops of food coloring to the glue and have the students use a toothpick to swirl the colors until they reach the desired effect. Let the glue dry and then gently peel their creations off of the lids. Use the hole punch to create a hole where the students can tie the string to hang up the suncatchers.
- **DAY-NIGHT COLLAGE** – Have the students make a collage depicting the things they do during the day vs. the things they do during the night. You can have them draw the pictures, cut out magazine pictures, or take and print out photos from their day for the collage.

Memorization

Copywork/Dictation
- **COPYWORK SENTENCE**
  As the Earth turns, day changes to night on one side of the globe. On the other side, night changes to day.
- **DICTATION SELECTION**
  The Earth rotates as it goes through its orbit around the sun. As Earth turns, day changes to night on one side of the globe. On the other side, night changes to day with the rotation of our planet. This is because, as Earth turns, different parts of the surface of the planet face the sun. We call this the day-night cycle, which takes about twenty-four hours to complete.

Chapter 10 Notes

Chapter 11: Grid Schedule

	Supplies Needed
Demo	• 8 Sandwich-style cookies, Picture of the phases of the moon (template on Appendix pg. 131)
Projects	• Black poster board, White toothpaste or shaving cream, Butter knife, Tape, Wiffle ball • 2 Sheets of paper, Scissors, Yellow and orange paint, Paintbrush

Chapter Summary

The chapter opens with the Sheen family, also known as the Shine-O-Mite crew, explaining why the twins were in danger and why they kidnapped them. The custodial crew convinces the twins to go to Paul Sims to get him to understand the threat that the Rotary Club poses to the museum. The three-member club plans to steal a guidance system from the museum to use in their rocket, which will destroy all communications satellites, ushering in a new era of rotary telephone use. The twins leave the Shine-O-Mite crew and meet back up with Summer and Paul in the Lunar Exploration Vehicles exhibit, where they learn about the moon. REESE sings them a song about lunar cycles and Summer dances off to find a good place for the asteroid she brought back. After that, the twins pull Paul to the side to tell him what they learned from the Shine-O-Mite crew. He dismisses their concerns and then shares about eclipses before heading off to find Summer. After he leaves, Blaine and Tracey overhear three people—Alexander, Graham, and Belle—discussing their plans for a heist, and the twins realize the they have just found the Rotary Club. They decide to go find Paul when a hand touches their shoulder. Paul has also overheard the three members and he a change of heart. The chapter ends with the twins, their local expert, and the Shine-O-Mite crew devising a plan to protect the museum and the guidance system from the impending heist.

Weekly Schedule

	Day 1	Day 2	Day 3	Day 4
Read	☐ Read the section entitled "The Lunar Rap" of Chapter 11 in *SSA Volume 6: Astronomy*.	☐ Read the section entitled "Eclipse Escapade" of Chapter 11 in *SSA Volume 6: Astronomy*.	☐ (*Optional*) Read one or all of the assigned pages from the encyclopedia of your choice.	☐ (*Optional*) Read one of the additional books from your library.
Write	☐ Fill out a Astronomy Record Sheet on SL pg. 51 on the moon. ☐ Add daily observations to the Moon Diary Sheet on SL pg. 48.	☐ Fill out a Astronomy Record Sheet on SL pg. 52 on eclipses. ☐ Go over the vocabulary word and enter it into the Astronomy Glossary on SL pg. 95. ☐ (*Optional*) Write a mission report.	☐ (*Optional*) Write narration on the Astronomy Notes Sheet on SL pg. 54. ☐ Add information learned from the demonstration on SL pg. 54.	☐ (*Optional*) Complete the copywork or dictation assignment and add it to the Astronomy Notes sheet on SL pg. 54. ☐ (*Optional*) Fill out the record sheet on SL pg. 56 for one of the projects. ☐ (*Optional*) Take Astronomy Quiz #5.
Do	☐ (*Optional*) Play a game of moon crater toss.	☐ (*Optional*) Make eclipse paintings.	☐ Do the demonstration entitled "Moon Cookies."	☐ Work on the Solar System Model.

Chapter 11: List Schedule

	Supplies Needed
Demo	• 8 Sandwich-style cookies, Picture of the phases of the moon (template on Appendix pg. 131)
Projects	• Black poster board, White toothpaste or shaving cream, Butter knife, Tape, Wiffle ball • 2 Sheets of paper, Scissors, Yellow and orange paint, Paintbrush

Chapter Summary

The chapter opens with the Sheen family, also known as the Shine-O-Mite crew, explaining why the twins were in danger and why they kidnapped them. The custodial crew convinces the twins to go to Paul Sims to get him to understand the threat that the Rotary Club poses to the museum. The three-member club plans to steal a guidance system from the museum to use in their rocket, which will destroy all communications satellites, ushering in a new era of rotary telephone use. The twins leave the Shine-O-Mite crew and meet back up with Summer and Paul in the Lunar Exploration Vehicles exhibit, where they learn about the moon. REESE sings them a song about lunar cycles and Summer dances off to find a good place for the asteroid she brought back. After that, the twins pull Paul to the side to tell him what they learned from the Shine-O-Mite crew. He dismisses their concerns and then shares about eclipses before heading off to find Summer. After he leaves, Blaine and Tracey overhear three people—Alexander, Graham, and Belle—discussing their plans for a heist, and the twins realize the they have just found the Rotary Club. They decide to go find Paul when a hand touches their shoulder. Paul has also overheard the three members and he a change of heart. The chapter ends with the twins, their local expert, and the Shine-O-Mite crew devising a plan to protect the museum and the guidance system from the impending heist.

Essential To-Do's

Read
- ☐ Read the section entitled "The Lunar Rap" of Chapter 11 in *SSA Volume 6: Astronomy*.
- ☐ Read the section entitled "Eclipse Escapade" of Chapter 11 in *SSA Volume 6: Astronomy*.

Write
- ☐ Fill out a Astronomy Record Sheet on SL pg. 51 on the moon.
- ☐ Add daily observations to the Moon Diary Sheet on SL pg. 48.
- ☐ Go over the vocabulary word and enter it into the Astronomy Glossary on SL pg. 95.
- ☐ Fill out a Astronomy Record Sheet on SL pg. 52 on eclipses.
- ☐ Add information learned from the demonstration on SL pg. 54.

Do
- ☐ Do the demo entitled "Moon Cookies."
- ☐ Work on the Solar System Model.

Optional Extras

Read
- ☐ Read one or all of the assigned pages from the encyclopedia of your choice.
- ☐ Read one of the additional books from your library.

Write
- ☐ Complete the copywork or dictation assignment and add it to the Astronomy Notes sheet on SL pg. 54.
- ☐ Fill out the record sheet on SL pg. 56 for one of the projects.
- ☐ Write a mission report.
- ☐ Take Astronomy Quiz #5.

Do
- ☐ Play a game of moon crater toss.
- ☐ Make eclipse paintings.

Chapter 11: A Ballistic Heist

Science-Oriented Books

Living Book Spine
- Chapter 11 of *The Sassafras Science Adventures Volume 6: Astronomy*

Optional Encyclopedia Readings
- *Basher Science Astronomy* pg. 22 (Moon)
- *Usborne Children's Encyclopedia* pp. 260-261 (The Moon)
- *DK First Space Encyclopedia* pp. 54-55 (Eclipse of the Sun), pp. 62-63 (The Moon)
- *Kingfisher Science Encyclopedia* pg. 401 (Earth and the Moon - Moon Section only), pg. 402 (Eclipses)

Additional Living Books
- *Faces of the Moon* by Bob Crelin and Leslie Evans
- *The Moon Book* by Gail Gibbons
- *The Moon Seems to Change (Let's-Read-and-Find Out Science 2)* by Franklyn M. Branley and Barbara and Ed Emberley
- *Eclipses (Amazing Sights of the Sky)* by Martha Elizabeth Hillman Rustad
- *Eclipse Chaser: Science in the Moon's Shadow (Scientists in the Field Series)* by Ilima Loomis and Amanda Cowan

Notebooking (SCIDAT Logbook Information)

This week, you can have the students complete a moon diary sheet. You can also have them fill out the logbook sheets for an eclipse and the moon. Here is the information they could include:

Moon Diary Sheet

This week, have the students record the phase of the moon each night by coloring in the portion of the moon that is visible on that day on the Moon Diary Sheet. The students will continue to keep a moon diary through chapter 14.

Astronomy Record Sheets
Eclipse
Information Learned

⇨ *An eclipse is when one celestial body obscures the light from another celestial body.*
⇨ *There are two types of eclipses we can see on Earth:*
  1. *Solar eclipse – These occur when the moon passes between the sun and the Earth, temporarily blocking out some of the light from the sun. The makes it seem like night in the middle of the day. Solar eclipses occur about every 15 months.*
  2. *Lunar eclipse – These occur when the Earth is between the sun and the moon, causing the moon to move into the Earth's shadow. The moon appears very dark and takes on a slight reddish glow. Lunar eclipses happen during a full moon about two times a year.*

⇨ *A total solar eclipse is when the sun is completely blocked out by the moon's shadow.*
⇨ *A partial solar eclipse is when it looks like a bite has been taken out of the sun because the moon's shadow only covers a portion.*

Moon

Information Learned

⇨ *The moon is our closest space neighbor.*
⇨ *It orbits the Earth in the same way that the Earth orbits the sun.*
⇨ *The moon does have an atmosphere, but it is composed of very different gases, such as sodium and potassium. There is no air that humans can breathe on the moon.*
⇨ *The surface of the moon is covered with craters that were made by rocks that crashed into it.*
⇨ *The moon takes about 27 days to go around the Earth and it takes about a month to spin. Because of this, on Earth we only see one side of the moon.*
⇨ *As the moon moves, parts of it are "lit" by the sun, which makes it look like the moon is changing shape.*
⇨ *The phases of the lunar cycle are:*
  1. Full Moon – During this phase, it appears as if the entire moon is lit.
  2. Gibbous Moon – During this phase, it appears as if about ¾ of the moon is lit.
  3. Last Quarter – During this phase, it appears as if half the moon is lit.
  4. Crescent Moon – During this phase, it appears as if ¼ of the moon is lit.
  5. New Moon – During this phase, it appears as if none of the moon is lit.
⇨ *The moon is said to be waxing if it is appearing to grow larger, e.g., moving from a waxing gibbous moon to a full moon, and waning if it is appearing to grow smaller, e.g., moving from a last quarter moon to a waning crescent.*

Vocabulary

Have the older students look up the following term in the glossary in the Appendix on pp. 137-138 or in a science encyclopedia. Then, have them copy each definition onto a blank index card or into their SCIDAT logbook.

☾ **Solar Eclipse** – The time when the moon blocks the sun.

Scientific Demonstration: Moon Cookies

Materials
☑ Sandwich-style cookies
☑ Picture of the phases of the moon (template on Appendix pg. 131)

Procedure
1. Have the students begin by opening one of the sandwich-style cookies. Have them use the have with icing for the full moon and the half without icing for the new moon.
2. Have the students continue to open up the cookies and remove a portion of the icing to match the different phases of the moon, using the picture you provided them as a guide.
3. When they are done, hold up one of the moon cookies and ask the students:
   ❓ Do you remember what this phase of the moon is?
4. Repeat with the other moon cookies until you have reviewed all of the phases.

Take it Further
Have the students make a paper-mâché model of the moon. You will need the following supplies: small balloon, newspaper, 1 cup of flour, ½ cup of water, 2 tablespoons of salt, paint, and a picture of the

moon. Have the students blow up the balloon. Next, have them tear the newspaper into strips. As they are working on the newspaper strips, use the flour, water, and salt to make a thick paste. You can add more or less water to gain the desired consistency. Then, have the students dip the strips into the paste mixture and cover the balloon with one layer. Wait 30 minutes before having them add a second and third layer if necessary. Finally, set the paper mâché moon in a place to dry. Once it is dry, have them use a picture of the moon as a guide to plaint their creations.

Multi-Week Projects and Activities

Multi-week Projects

- **Solar System Model** – This week, the students will add Earth's moon to their model. Have the students cut out the moon from copy paper or construction paper using the diameter below. (NOTE—*The moon is one-fourth the size of the Earth, so this might be a bit of a challenge for the lap-sized version. If your students are creating a lap-sized version, you can have them just draw the moon directly on their model near the Earth rather than cutting it out.*) Have them color the moon using a picture from the suggested books or from the Internet. When they are done, have them add it to their solar system just above their Earth.
  - ⇨ Diameter (wall version): 0.5 in
  - ⇨ Distance (lap version): 0.25 cm

Activities For This Week

- **Moon Crater Toss** – Have the students play a game of moon crater toss. You will need a sheet of black poster board, white toothpaste or shaving cream, a butter knife, tape, and a Wiffle ball. Directions for this game can be found at the following website:
  - https://www.pre-kpages.com/moon-crater-gross-motor-activity/
- **Eclipse Painting** – Have the students paint the sun's corona, which is visible during a solar eclipse. You will need 2 sheets of paper, scissors, yellow and orange paint, and a paintbrush. Cut out a circle from one of the sheets of paper. Hold the circle in the middle of the paper and have the students use the yellow and orange paint to away and off the edge of the circle to form the sun's corona. After they are done, remove the circle. See what the finished project looks like here:
  - https://www.instagram.com/p/BYEutQWHg5T/

  You can also have the students add a poem or narration about eclipses to the center of their paintings.
- **Mission Report** – Have the students research and write a brief report about one of the most famous journey to the moon - *Apollo 11*. They should include the dates, several pictures, and a summary of what was found in their report. (NOTE—*The students will study more about Neil Armstrong and his moon walk during the last week of this program.*) You can watch a video of the very first moon walk here:
  - https://www.youtube.com/watch?v=RMINSD7MmT4

Memorization

Copywork/Dictation

- **Copywork Sentence**

  A total solar eclipse is when the sun is completely blocked out by the moon's shadow.

- **Dictation Selection**

  An eclipse is when one celestial body obscures the light from another celestial body. There are two types of eclipses we can see on Earth: solar eclipses and lunar eclipses. A solar eclipse occurs when the moon passes between the sun and the Earth, temporarily blocking out some of the light from the sun. A lunar eclipse occurs when the Earth is between the sun and the moon, causing the moon to move into the Earth's shadow.

Quiz Information

This week, you can give the students a quiz based on what they learned in chapters 10 and 11. You can find this quiz in the Appendix on pg. 149.

Quiz #5 Answers
1. Star
2. True
3. Towards, Away from
4. False (*A full day-night cycle takes about 24 hours.*)
5. Solar
6. Lunar
7. True
8. New moon, Full moon

Chapter 11 Notes

Chapter 12: Grid Schedule

	Supplies Needed
Demo	• Thick rubber gloves or work gloves, LEGO bricks, Several bolts, washers, and nuts that fit each other
Projects	• Soda bottle, Cardstock, White and black paint, Glue • String, Small paper cup, Mini-marshmallows, Potential parachute material (paper, tissue, thin fabric, felt), Scissors or a hole punch

Chapter Summary

The chapter opens with the museum heist by the Rotary Club in progress. Tracey and REESE manage to remove the guidance component to keep it safe, and after a few tense moments the Rotary Club is caught. Meanwhile, we find out that Adrianna Archer has gone to Uncle Cecil's house. The next day, Blaine and Tracey open LINLOC for their next location, but something is wrong. The twins are able to see a few of the subjects they need info on, but the app is definitely not working. We find out that the error is probably due to Adrianna Archer smashing Uncle Cecil's keyboard. Back at the museum, the twins are learning about astronauts from Paul Sims while Summer calls Cecil. Adrianna Archer answers Cecil's phone and an interesting conversation ensues. Once done, Summer rejoins the twins as they are learning about space shuttles from Paul. They are all confused by the woman who answered Cecil's phone. As they reason it out, we switch back to Uncle Cecil, who promptly kicks Adrianna out of his house and manages to fix the LINLOC app. Back at the museum, the twins decide to try LINLOC one more time, and this time it works. The twins zip off to Poland and land in a very strange, darkened room with a pendulum swinging in the center. The chapter ends with Blaine and Tracey finding out that they are about to take part in the "Copernicus Code Escape Room."

Weekly Schedule

	Day 1	Day 2	Day 3	Day 4
Read	☐ Read the section entitled "Abort the Astronaut Adventure" of Chapter 12 in *SSA Volume 6: Astronomy*.	☐ Read the section entitled "Space Shuttle Stalemate" of Chapter 12 in *SSA Volume 6: Astronomy*.	☐ (*Optional*) Read one or all of the assigned pages from the encyclopedia of your choice.	☐ (*Optional*) Read one of the additional books from your library.
Write	☐ Fill out a Astronomy Record Sheet on SL pg. 59 on astronauts. ☐ Add daily observations to the Moon Diary Sheet on SL pg. 57.	☐ Fill out a Astronomy Record Sheet on SL pg. 60 on space shuttles. ☐ Go over the vocabulary word and enter it into the Astronomy Glossary on SL pg. 95.	☐ (*Optional*) Write narration on the Astronomy Notes Sheet on SL pg. 63. ☐ Add information learned from the demonstration on SL pg. 63.	☐ (*Optional*) Complete the copywork or dictation assignment and add it to the Astronomy Notes sheet on SL pg. 63. ☐ (*Optional*) Fill out the record sheet on SL pg. 65 for one of the projects.
Do	☐ (*Optional*) Do a bit of astronaut training.	☐ (*Optional*) Make a model shuttle.	☐ Do the demonstration entitled "Space Tasks."	☐ Do the astronaut drop challenge.

Chapter 12: List Schedule

	Supplies Needed
Demo	• Thick rubber gloves or work gloves, LEGO bricks, Several bolts, washers, and nuts that fit each other
Projects	• Soda bottle, Cardstock, White and black paint, Glue • String, Small paper cup, Mini-marshmallows, Potential parachute material (paper, tissue, thin fabric, felt), Scissors or a hole punch

Chapter Summary

The chapter opens with the museum heist by the Rotary Club in progress. Tracey and REESE manage to remove the guidance component to keep it safe, and after a few tense moments the Rotary Club is caught. Meanwhile, we find out that Adrianna Archer has gone to Uncle Cecil's house. The next day, Blaine and Tracey open LINLOC for their next location, but something is wrong. The twins are able to see a few of the subjects they need info on, but the app is definitely not working. We find out that the error is probably due to Adrianna Archer smashing Uncle Cecil's keyboard. Back at the museum, the twins are learning about astronauts from Paul Sims while Summer calls Cecil. Adrianna Archer answers Cecil's phone and an interesting conversation ensues. Once done, Summer rejoins the twins as they are learning about space shuttles from Paul. They are all confused by the woman who answered Cecil's phone. As they reason it out, we switch back to Uncle Cecil, who promptly kicks Adrianna out of his house and manages to fix the LINLOC app. Back at the museum, the twins decide to try LINLOC one more time, and this time it works. The twins zip off to Poland and land in a very strange, darkened room with a pendulum swinging in the center. The chapter ends with Blaine and Tracey finding out that they are about to take part in the "Copernicus Code Escape Room."

Essential To-Do's

Read
☐ Read the section entitled "Abort the Astronaut Adventure" of Chapter 12 in *SSA Volume 6: Astronomy*.
☐ Read the section entitled "Space Shuttle Stalemate" of Chapter 12 in *SSA Volume 6: Astronomy*.

Write
☐ Add daily observations to the Moon Diary Sheet on SL pg. 57.
☐ Fill out a Astronomy Record Sheet on SL pg. 59 on astronauts.
☐ Go over the vocabulary word and enter it into the Astronomy Glossary on SL pg. 95.
☐ Fill out a Astronomy Record Sheet on SL pg. 60 on space shuttles.
☐ Add information learned from the demonstration on SL pg. 63.

Do
☐ Do the demonstration entitled "Space Tasks."
☐ Do the astronaut drop challenge.

Optional Extras

Read
☐ Read one or all of the assigned pages from the encyclopedia of your choice.
☐ Read one of the additional books from your library.

Write
☐ Write a narration on the Astronomy Notes Sheet on SL pg. 63.
☐ Fill out the record sheet on SL pg. 65 for one of the projects.
☐ Complete the copywork or dictation assignment and add it to the Astronomy Notes sheet on SL pg. 63.

Do
☐ Do a bit of astronaut training.   ☐ Make a model shuttle.

Chapter 12: LINLOC Failure to Launch

Science-Oriented Books

Living Book Spine
- Chapter 12 of *The Sassafras Science Adventures Volume 6: Astronomy*

Optional Encyclopedia Readings
- *Basher Science Astronomy* (No pages scheduled)
- *Usborne Children's Encyclopedia* pp. 248-249 (Trips into Space)
- *DK First Space Encyclopedia* pp. 26-27 (Astronaut in Training), pp. 36-37 (Space Shuttle)
- *Kingfisher Science Encyclopedia* pg. 421 (Rockets and Spaceplanes - Section on the space shuttle), pg. 422 (Humans in Space)

Additional Living Books
- *Mousetronaut: Based on a (Partially) True Story (Paula Wiseman Books)* by Mark Kelly and C. F. Payne
- *If You Decide to Go to the Moon* by Faith McNulty and Steven Kellogg
- *DK Readers L2: Astronaut: Living in Space* by Deborah Lock
- *Floating in Space (Let's-Read-and-Find-Out Science 2)* by Franklyn M. Branley and True Kelley
- *DK Readers L1: Rockets and Spaceships* by Karen Wallace

Notebooking (SCIDAT Logbook Information)

This week, you can have the students complete a moon diary sheet. You can also have them fill out the logbook sheets for astronauts and the space shuttle. Here is the information they could include:

Moon Diary Sheet
This week, have the students record the phase of the moon each night by coloring in the portion of the moon that is visible on that day on the Moon Diary Sheet. The students will continue to keep a moon diary through chapter 14.

Astronomy Record Sheets
Astronauts
Information Learned
- Astronauts must wear a space suit, kind of like a personalized spacecraft, in order to work out in space.
- The space suit has an outfit with tubes throughout. These tubes are filled with water that can help to quickly cool or heat the astronaut.
- The outer layers of the space suit help to protect the astronaut from meteoroids and from the extreme temperatures in space.
- The astronaut also wears a helmet fitted with a radio, drinking tube, and visor to protect the astronaut from the sun.
- The astronaut wears padded gloves with rubber tips, so he can feel and grip things easily.
- Finally, the astronaut puts on a "Primary Life Support System," which contains the air he needs to breathe and helps to monitor and regulate the astronaut's temperature.

- *The whole space suit weighs about 310 pounds!*
- *An astronaut trains for years to handle space. They have to learn how to live in zero gravity, how to escape the space shuttle quickly, and to work in space in a space suit.*
- *Astronauts train to do tasks in the space suit in giant pools, since floating in water is slightly similar to floating in space.*

Space Shuttle
Information Learned

- *The space shuttle is reusable.*
- *The space shuttle has two boosters, a large fuel tank, and the main reusable shuttle.*
- *When the space shuttle is launched, the shuttle blasts off a launch pad.*
- *The booster rockets use up their fuel first, fall away, and parachute back to earth.*
- *Then, the large fuel tank powers the main engine and falls away when it runs out of fuel, which happens once the shuttle has been delivered into space.*
- *The space shuttle has compartments to hold the crew and fly the shuttle in the front.*
- *In the middle is a large cargo bay that can carry telescopes, satellites, or parts for a space station.*
- *The wings on the shuttle help to guide the aircraft back to the earth.*

Vocabulary

Have the older students look up the following term in the glossary in the Appendix on pp. 137-138 or in a science encyclopedia. Then, have them copy each definition onto a blank index card or into their SCIDAT logbook.

- **ASTRONAUT** – A person who travels into space.

Scientific Demonstration: Space Tasks

Materials
- ☑ Thick yellow rubber gloves or work gloves
- ☑ LEGO bricks
- ☑ Several bolts, washers, and nuts that fit each other

Procedure
1. Place the LEGO bricks along with the screws, washers, and bolts, on the table in front of the students.
2. Have the students build a tower with the LEGO bricks. After that, have them put a washer on each bolt and secure it with a nut.
3. Have the students put on the thick, rubber gloves as you take apart the work they did in the previous step. (**NOTE**—*For an added challenge, you can have the students also use knit glove liners under the rubber gloves.*)
4. Now have them attempt to build the same tower with the LEGO bricks. After that, have them attempt to put a washer on each bolt and secure it with a nut once more.
   ? Was it different with the gloves?

Explanation
The students should see that the gloves made all the tasks more difficult. The same is true for the astronauts; the space suit, complete with thick gloves, makes every task they do more difficult. Add in a lack of gravity and you can see why it requires so much training to be an astronaut!

Take it further
Repeat the process from the demonstration, only this time use a glove box for an added challenge. You can see how to built your own glove box here:
- https://www.giftofcuriosity.com/diy-astronaut-glove-box/

Multi-Week Projects and Activities

Multi-week Projects
- **Solar System Model** - There is nothing to add to the solar system model this week.

Activities for This Week
- **Astronaut Training** – Have the students practice training like an astronaut. NASA has a whole list of exercises that astronauts use as a part of their training program. You can view them all here:
  - http://www.nasa.gov/audience/foreducators/trainlikeanastronaut/home/
- **Model Shuttle** – Have the students make a model space shuttle. You will need a soda bottle, cardstock, white and black paint, and glue. The directions for this project can be found at the following website:
  - http://www.daniellesplace.com/html/outer-space-crafts.html
- **Astronaut Drop Challenge** – Have the students test to see if they can create a parachute with enough drag to slow down a paper cup space capsule filled with astronaut marshmallows. You will need string, a small paper cup, mini-marshmallows, potential parachute material (paper, tissue, thin fabric, felt), and scissors or a hole punch. Have the students use the material to design a parachute. You can show them pictures of parachutes beforehand or read about the principles of flight using the following:
  - *Usborne Children's Encyclopedia* pp. 240-241 (Planes)
  - *Kingfisher Science Encyclopedia* pp. 332-333 (Principles of Flight)

  Or you can just let their imaginations run wild! After they have their parachutes, have the students punch two holes in the top of the paper cup and use string to attach the parachute they design. Add 4 to 5 mini-marshmallows to the cup and then hold it over your head and drop to see what happens to the astronaut marshmallows! If the "astronauts" don't survive and you have time, let the students design another parachute to see if they get better results.

Memorization

Copywork/Dictation
- **Copywork Sentence**

  Astronauts are people who travel into space and perform tasks while there.

- **Dictation Selection**

  The first person went into space in April 1961. The flight took Soviet astronaut, Yuri Gagarin, into space and around the Earth one time. The first American astronaut to enter space was Alan Shepard in May of 1961. The early trips into space were done via rockets that were not reusable. Then, the reusable space shuttle was designed.

Chapter 12 Notes

Chapter 13: Grid Schedule

	Supplies Needed
Demo	• Thin wooden dowel or a straw, String (about 12" long), Heavy metal nut or washer, Protractor, Tape
Projects	• 3' Curling ribbon, Tennis ball, Foil, Straight pin

Chapter Summary

The chapter opens with the twins learning that they are going to have to escape the Copernicus Code Room before the pendulum knocks down all of the blocks in the circle. Minka explains that the room was named after the famous astronomer Nicholas Copernicus. She tells them that she will give them clues to help them along the way. The twins find out that they are not alone; they are joined by a brother and sister named Clive and Halley. The first clue is all about comets and after the four solve the riddle, they find a quadrant. Minka explains this is their second clue and it also involves the ancient astronomer Ptolemy. Once more, they work out the clue and learn about another ancient astronomer, Nicholas Copernicus. The four children figure out the clue and open a hole to another section of the escape room. The next clue involves Johannes Kepler, which they quickly work out and unlock a door into yet another part of the escape room. They enter a mirror maze and get another clue, this one about Galileo. The work their way through to a set of stairs. Minka announces that all they have to do is climb the stairs and capture the light, but they have to do it quickly because there are only two blocks remaining. The chapter ends with the four thinking they have figured out how to capture the light and jumping, hopeful they have escaped in time.

Weekly Schedule

	Day 1	**Day 2**	**Day 3**	**Day 4**
Read	☐ Read the section entitled "Comet Circumvention" of Chapter 13 in *SSA Volume 6: Astronomy*.	☐ Read the section entitled "AWOL Ancients" of Chapter 13 in *SSA Volume 6: Astronomy*.	☐ *(Optional)* Read one or all of the assigned pages from the encyclopedia of your choice.	☐ *(Optional)* Read one of the additional books from your library.
Write	☐ Fill out a Astronomy Record Sheet on SL pg. 61 on comets. ☐ Add information about Ptolemy and Galileo to SL pg. 62. ☐ Add daily observations to the Moon Diary Sheet on SL pg. 58.	☐ Fill out a Astronomy Record Sheet on SL pg. 62 on the remaining ancient astronomers. ☐ Go over the vocabulary word and enter it into the Astronomy Glossary on SL pg. 96.	☐ *(Optional)* Write narration on the Astronomy Notes Sheet on SL pg. 64. ☐ Add information learned from the demonstration on SL pg. 64.	☐ *(Optional)* Complete the copywork or dictation assignment and add it to the Astronomy Notes sheet on SL pg. 64. ☐ *(Optional)* Fill out the record sheet on SL pg. __ for one of the projects. ☐ *(Optional)* Take Astronomy Quiz #6.
Do	☐ *(Optional)* Make the flying comet model.	☐ *(Optional)* Begin the scientist biography report.	☐ Do the demonstration entitled "Simple Astrolabe."	☐ Work on the Solar System Model. ☐ *(Optional)* Finish the scientist biography report.

Chapter 13: List Schedule

	Supplies Needed
Demo	• Thin wooden dowel or a straw, String (about 12" long), Heavy metal nut or washer, Protractor, Tape
Projects	• 3' Curling ribbon, Tennis ball, Foil, Straight pin

Chapter Summary

The chapter opens with the twins learning that they are going to have to escape the Copernicus Code Room before the pendulum knocks down all of the blocks in the circle. Minka explains that the room was named after the famous astronomer Nicholas Copernicus. She tells them that she will give them clues to help them along the way. The twins find out that they are not alone; they are joined by a brother and sister named Clive and Halley. The first clue is all about comets and after the four solve the riddle, they find a quadrant. Minka explains this is their second clue and it also involves the ancient astronomer Ptolemy. Once more, they work out the clue and learn about another ancient astronomer, Nicholas Copernicus. The four children figure out the clue and open a hole to another section of the escape room. The next clue involves Johannes Kepler, which they quickly work out and unlock a door into yet another part of the escape room. They enter a mirror maze and get another clue, this one about Galileo. The work their way through to a set of stairs. Minka announces that all they have to do is climb the stairs and capture the light, but they have to do it quickly because there are only two blocks remaining. The chapter ends with the four thinking they have figured out how to capture the light and jumping, hopeful they have escaped in time.

Essential To-Do's

Read
☐ Read the section entitled "Comet Circumvention" of Chapter 13 in *SSA Volume 6: Astronomy*.
☐ Read the section entitled "AWOL Ancients" of Chapter 13 in *SSA Volume 6: Astronomy*.

Write
☐ Add daily observations to the Moon Diary Sheet on SL pg. 58.
☐ Fill out a Astronomy Record Sheet on SL pg. 61 on comets.
☐ Go over the vocabulary word and enter it into the Astronomy Glossary on SL pg. 96.
☐ Fill out a Astronomy Record Sheet on SL pg. 62 on ancient astronomers.
☐ Add information learned from the demonstration on SL pg. 64.

Do
☐ Do the demo entitled "Simple Astrolabe."
☐ Work on the Solar System Model.

Optional Extras

Read
☐ Read one or all of the assigned pages from the encyclopedia of your choice.
☐ Read one of the additional books from your library.

Write
☐ Write a narration on the Astronomy Notes Sheet on SL pg. 64.
☐ Complete the copywork or dictation assignment and add it to the Astronomy Notes sheet on SL pg. 64.
☐ Fill out the record sheet on SL pg. 66 for one of the projects.
☐ Take Astronomy Quiz #6.

Do
☐ Make the flying comet model.
☐ Do the scientist biography report.

Chapter 13: The Copernicus Code

Science-Oriented Books

Living Book Spine
- Chapter 13 of *The Sassafras Science Adventures Volume 6: Astronomy*

Optional Encyclopedia Readings
- *Basher Science Astronomy* pg. 4 (Galileo), pg. 54 (Halley's Comet)
- *Usborne Children's Encyclopedia* pg. 275 (Comets)
- *DK First Space Encyclopedia* pp. 8-9 (Stargazers), pp. 76-77 (Comets and Meteors)
- *Kingfisher Science Encyclopedia* pg. 412 (Comets), pp. 414-415 (Studying the Universe)

Additional Living Books
- *Comets, Meteors, and Asteroids* by Seymour Simon
- *Asteroids, Comets, and Meteorites (First Facts)* by Steve Kortenkamp
- *Asteroids and Comets (Science Readers: A Closer Look)* by William B. Rice
- *Comets and Asteroids: Space Rocks (Our Solar System)* by Greg Roza

Notebooking (SCIDAT Logbook Information)

This week, you can have the students complete a moon diary sheet. You can also have them fill out the logbook sheets for comets and ancient astronomers. Here is the information they could include:

Moon Diary Sheet
This week, have the students record the phase of the moon each night by coloring in the portion of the moon that is visible on that day on the Moon Diary Sheet. The students will continue to keep a moon diary through chapter 14.

Astronomy Record Sheets
Comets
Information Learned
- A comet is like a giant space snowball. They are named after their description in Greek "aster kometes," which means "hairy star."
- Comets are made from dust and ice. They come from the outer edge of our solar system and orbit the sun, traveling between the planets.
- As a comet passes Jupiter and gets closer to the sun, the ice starts to melt causing clouds of dust and gas. This cloud, called a comas, forms around the comet, forming a visible tail that seems to glow.
- A comet's tail always points away from the sun. This is because the gases and dust particles are being pushed by solar wind.
- In 1997, the comet Hale-Bopp came close enough to the Earth for us to see it without a telescope, tail and all, clearly. It won't return near to Earth for another 2,300 years.
- Halley's Comet is the most famous returning comet that is visible from Earth. This comet returns about every 75 years and is set to be seen in our skies again in 2061. It is named for the scientist, Edmond

Halley, who predicted its return in 1705. He noticed that there were similar descriptions of a comet being seen in the sky recorded at regular intervals. He used Isaac Newton's recent discoveries of gravity and motion to predict when the comet would be seen again. He didn't live to see it, but sure enough, his comet appeared in the sky once again in 1758.

Ancient Astronomers
Information Learned

- *Ptolemy* – Claudius Ptolemy was an Egyptian with Greek heritage who studied math, the Earth, and the stars. He was alive during the 2nd century, but we don't know exactly when as not much about his life is known. We do know that he wrote several scientific books, including one that detailed his work on cataloguing the stars and the positions of the planets. He believed that all these things orbited around the Earth. His work was the astronomical authority for thousands of years. Even today, 48 of our constellations still bear the names Ptolemy gave them.

- *Copernicus* – Nicholas Copernicus, who lived during the 15th and 16th centuries, changed everything the world knew about astronomy when he proposed that the sun, not the Earth, was at the center of our universe. He said that the planets moved in a uniform path around the sun, instead of the irregular path around Earth that Ptolemy had suggested. His ideas were not officially adopted until about a hundred years after his death, but his mark on astronomy is still felt today.
- *Kepler* – Johannes Kepler, who living during the 16th and 17th centuries, build on Copernicus's work. He agreed that the planets orbit the sun but said that they do so in an elliptical path rather than the perfect circle Copernicus suggested. He did a lot of significant work in astronomy on math and optics, but he is best known for his three laws that explain the relationship between a planet's distance from the sun and the length of its orbit.
- *Galileo* – Galileo Galilei, who also living during the 16th and 17th centuries, used the newly invented telescope to probe that Copernicus was correct – the Earth and the other planets did orbit around the sun. He also discovered Jupiter's moons and had a few ideas about falling objects that Isaac Newton later built upon as he wrote his theory of gravity and the Three Laws of Motion. Galileo also invented the first telescopes, capable of magnifying things by 20 times.

Vocabulary

Have the older students look up the following terms in the glossary in the Appendix on pp. 127-128 or in a science encyclopedia. Then, have them copy each definition onto a blank index card or into their SCIDAT logbook.

- ASTRONOMER – A person who studies space and the things found in it.

Scientific Demonstration: Simple Astrolabe

Materials
- ☑ Thin wooden dowel or a straw
- ☑ String (about 12" long)
- ☑ Heavy metal nut or washer
- ☑ Protractor
- ☑ Tape

Procedure
1. Have the students tie a string to the middle of the wooden dowel, or straw, and tape it so that it stays

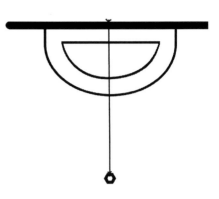

in place. Then, have them tie the metal nut to the other end of the string.

2. Tape the protractor onto the wooden dowel, or straw, so that the middle of the long straight end is at the same place as the string. (*See the diagram for help in visualizing the set-up.*)
3. Have the students close one eye and then use the other to look down the line of the top of the wooden dowel, or through the hole in the straw, taking caution to not poke their eye.
4. Have the students line up the wooden dowel, or straw, with the top of an object in the distance and have a partner note the angle of the hanging string on the protractor.
5. Have the students repeat the process for several other distant objects.

Explanation
The students should see that the angle of the hanging string increases as the height of the objects increase. This simple principle was used by ancient astronomers to measure the distance between stars because as the distance increased, the angle also increased.

Take it further
Have the students read more about astrolabes and what they were used for throughout history in this article from Encyclopedia Britannica:
- https://www.britannica.com/science/astrolabe-instrument

Multi-Week Projects and Activities

Multi-week Projects
- **Solar System Model** - Have the students add a comet to their solar system model. They should add the comet with a tail pointing away from the sun anywhere before Jupiter.

Activities For This Week
- **Scientist Biography** – Have the students read a biography on a famous astronomer. We recommend that you choose one of the four astronomers mentioned in the novel, but you can also have your students choose a different astronomer. To learn about the scientist, you can get book from the library or have the students do a bit of research on the individual on the Internet. You can read the book out loud to your students, or have them read it on their own. Either way, as they read have the students answer the following questions about the scientist.
    - ? Who was the scientist you read about?
    - ? When and where were he born?
    - ? What was his major scientific contribution?
    - ? List the events that surround his discovery.
    - ? List some other interesting events in the scientist's life.
    - ? Why do you think that it is important to learn about this scientist?

  You can find scientist biography questionnaire on pp. 132-133 of this guide to give to the students.
- **Model Comet** – Have the students make their own comet that flies. You will need about three feet of curling ribbon, a tennis ball, foil, and a straight pin. Have them cut the curling ribbon into lengths, curl them, and tie them together at one end. Next, have the students pin the tied end of the ribbon onto the tennis ball using the straight pin. Then, cut out a piece of foil, about 7" x 7", for them to use to cover the ball. Make sure they allow space for the comet's tail (ribbon) to come out through the foil. Once they are done, they can have their comets fly through outer space.

Memorization

Copywork/Dictation

☞ **Copywork Sentence**

A comet is like a giant space snowball. It is made from dust and ice.

☞ **Dictation Selection**

Comets come from the outer edge of our solar system and orbit the sun. As a comet passes Jupiter and gets closer to the sun, the ice starts to melt causing clouds of dust and gas. This cloud, called a comas, forms around the comet, forming a visible tail that seems to glow. A comet's tail always points away from the sun because the gases and dust particles are being pushed by solar wind.

Quiz Information

This week, you can give the students a quiz based on what they learned in chapters 12 and 13. You can find this quiz in the Appendix on pg. 151.

Quiz #6 Answers

1. Underwater
2. Into space
3. True
4. B, A, C
5. False (*Comets are made of dust and ice.*)
6. Towards
7. B, A, D, C

Chapter 13 Notes

Chapter 14: Grid Schedule

	Supplies Needed
Demo	• Straw, String (5 feet), Scissors, Large balloon, 2 Chairs, Tape
Projects	• Build-a-rocket kit

Chapter Summary

The chapter opens with the twins having escaped the Copernicus Code Room, ready to zip off to their next location in New Zealand. Luckily, LINLOC has not been messed up by the fact that they got their SCIDAT information from two different experts. The twins arrive and are quickly ordered to get back to work as space walkers. After a bit of confusion, the twins are directed to the lead foreman, and their local expert, Arty Stone. He shares about space walks and explains how their job as space walkers at the New Zealand Space Games differs from that. They learn about the games and about the three different versions of Planet Prowess that they gamers will play. The twins listen to the announcer, Mr. Sebastian, introduces the gamers that will be participating – Wayne Hammer, Ms. Pink Rocker, Mohawk Wellington, Agnes the Librarian, and *El Cohete Loco*, otherwise known as The Crazy Rocket. After that, Arty shares about rockets before the twins learn about the legendary-but-mysterious player Robbie Thistler. They learn about his disappearance and the speculation that he will return this year to play the all-new virtual-reality edition of Planet Prowess. We flip to Pecan Street where the Man With No Eyebrows is searching for Adrianna Archer. The chapter ends back in New Zealand with the twins discovering that someone has finally beaten Robbie Thistler's record.

Weekly Schedule

	Day 1	Day 2	Day 3	Day 4
Read	☐ Read the section entitled "Sporty Space Walk" of Chapter 14 in *SSA Volume 6: Astronomy*.	☐ Read the section entitled "Rocket Recreations" of Chapter 14 in *SSA Volume 6: Astronomy*.	☐ (*Optional*) Read one or all of the assigned pages from the encyclopedia of your choice.	☐ (*Optional*) Read one of the additional books from your library.
Write	☐ Fill out a Astronomy Record Sheet on SL pg. 69 on space walks. ☐ Add daily observations to the Moon Diary Sheet on SL pg. 67.	☐ Fill out a Astronomy Record Sheet on SL pg. 70 on rockets. ☐ Go over the vocabulary words and enter them into the Astronomy Glossary on SL pg. 96.	☐ (*Optional*) Write narration on the Astronomy Notes Sheet on SL pg. 73. ☐ Add information learned from the demonstration on SL pg. 73.	☐ (*Optional*) Complete the copywork or dictation assignment and add it to the Astronomy Notes sheet on SL pg. 73. ☐ (*Optional*) Fill out the record sheet on SL pg. 75 for one of the projects.
Do	☐ (*Optional*) Watch the laws of motion video.	☐ (*Optional*) Build your rocket.	☐ Do the demonstration entitled "Balloon Rocket."	☐ (*Optional*) Launch your rocket.

Chapter 14: List schedule

Supplies Needed	
Demo	• Straw, String (5 feet), Scissors, Large balloon, 2 Chairs, Tape
Projects	• Build-a-rocket kit

Chapter Summary

The chapter opens with the twins having escaped the Copernicus Code Room, ready to zip off to their next location in New Zealand. Luckily, LINLOC has not been messed up by the fact that they got their SCIDAT information from two different experts. The twins arrive and are quickly ordered to get back to work as space walkers. After a bit of confusion, the twins are directed to the lead foreman, and their local expert, Arty Stone. He shares about space walks and explains how their job as space walkers at the New Zealand Space Games differs from that. They learn about the games and about the three different versions of Planet Prowess that they gamers will play. The twins listen to the announcer, Mr. Sebastian, introduces the gamers that will be participating – Wayne Hammer, Ms. Pink Rocker, Mohawk Wellington, Agnes the Librarian, and *El Cohete Loco*, otherwise known as The Crazy Rocket. After that, Arty shares about rockets before the twins learn about the legendary-but-mysterious player Robbie Thistler. They learn about his disappearance and the speculation that he will return this year to play the all-new virtual-reality edition of Planet Prowess. We flip to Pecan Street where the Man With No Eyebrows is searching for Adrianna Archer. The chapter ends back in New Zealand with the twins discovering that someone has finally beaten Robbie Thistler's record.

Essential To-Do's

Read
☐ Read the section entitled "Sporty Space Walk" of Chapter 14 in *SSA Volume 6: Astronomy*.
☐ Read the section entitled "Rocket Recreations" of Chapter 14 in *SSA Volume 6: Astronomy*.

Write
☐ Add daily observations to the Moon Diary Sheet on SL pg. 67.
☐ Fill out a Astronomy Record Sheet on SL pg. 69 on space walks.
☐ Go over the vocabulary words and enter them into the Astronomy Glossary on SL pg. 96.
☐ Fill out a Astronomy Record Sheet on SL pg. 70 on rockets.
☐ Add information learned from the demonstration on SL pg. 73.

Do
☐ Do the demonstration entitled "Balloon Rocket."

Optional Extras

Read
☐ Read one or all of the assigned pages from the encyclopedia of your choice.
☐ Read one of the additional books from your library.

Write
☐ Write a narration on the Astronomy Notes Sheet on SL pg. 73.
☐ Complete the copywork or dictation assignment and add it to the Astronomy Notes sheet on SL pg. 73.
☐ Fill out the record sheet on SL pg. 75 for one of the projects.

Do
☐ Watch a video on the Laws of Motion.
☐ Build and launch a rocket.

Chapter 14: The New Zealand Space Games

Science-Oriented Books

Living Book Spine
- Chapter 14 of *The Sassafras Science Adventures Volume 6: Astronomy*

Optional Encyclopedia Readings
- *Basher Science Astronomy* (No pages scheduled)
- *Usborne Children's Encyclopedia* pp. 194-195 (Forces)
- *DK First Space Encyclopedia* pp. 30-31 (Rockets), pp. 42-43 (Working in Space)
- *Kingfisher Science Encyclopedia* pp. 420-421 (Rockets and Spaceplanes)

Additional Living Books
- *Spacewalks (Little Astronauts)* by Kathryn Clay
- *Space Walks (Our Solar System)* by Dana Meachen Rau and Nadia Higgins
- *Endurance, Young Readers Edition: My Year in Space and How I Got There Paperback* by Scott Kelly
- *DK Readers L2: Spaceships and Rockets: Relive Missions to Space (DK Readers Level 2)* by DK
- *Go for the Moon: A Rocket, a Boy, and the First Moon Landing* by Chris Gall

Notebooking (SCIDAT Logbook Information)

This week, you can have the students complete a moon diary sheet. You can also have them fill out the logbook sheets for spacewalks and rockets. Here is the information they could include:

Moon Diary Sheet
This week, have the students record the phase of the moon each night by coloring in the portion of the moon that is visible on that day on the Moon Diary Sheet. This is the last week the students will add to their moon diary.

Astronomy Record Sheets
Space walks
Information Learned
- A space walk is a period of time spend outside of a spacecraft by an astronaut.
- Astronauts will do space walks to repair man-made satellites, to check on the outside of their spacecrafts, or to help build parts of the International Space Station.
- Astronauts have to use their special space suit to survive when they take a walk in space and the suit makes even the simple task more difficult to do.
- When the space walk faces the sun, it is boiling hot, but as soon as they travel into the earth's shadow, the temperature plummets way below freezing.
- The first space walk was done by Alexi Leonov on March 18th, 1965.
- Shortly after that on June 3rd, 1965, Ed White became the first American to complete a space walk.
- On July 12th, 1969 Neil Armstrong and Buzz Aldrin completed one of the most famous space walks, where they actually walked for the first time on the moon.
- Fewer than 600 people have been in space and only 12 of those people have walked on the moon.

Rockets

Information Learned

⇨ Rockets are responsible for carrying things and people into space.
⇨ Rockets burn liquid fuel in order to produce a stream of hot gas to propel the rocket up.
⇨ The first rocket was launched by Robert Goddard in 1926. It reached a height of 41 feet and the flight lasted 2.5 seconds.
⇨ There are many types of rockets, including reusable space shuttles, firework rockets, military rockets, and experimental rockets.
⇨ The largest and most powerful rockets build were the Saturn V rockets. They were used between 1968 and 1972 13 times, including the first moon landing.
⇨ A rocket's payload, or the cargo it is carrying, is what determines what size the rocket needs to be.
⇨ To escape Earth's gravity, the rocket has to reach a speed over 7 miles per second, which is known as the escape velocity.

Vocabulary

Have the older students look up the following terms in the glossary in the Appendix on pp. 137-138 or in a science encyclopedia. Then, have them copy each definition onto a blank index card or into their SCIDAT logbook.

- ESCAPE VELOCITY – The speed an object must reach in order to escape the force of Earth's gravity.
- FORCE – A push or pull on an object.

Scientific Demonstration: Balloon rocket

Materials

☑ Straw
☑ String (5 feet)
☑ Scissors
☑ Large balloon
☑ 2 Chairs
☑ Tape

Procedure

1. Have the students cut the straw in half and thread the string through the straw.
2. As they are doing this, set the chairs up so that the backs of the chairs are facing each other about 5 feet apart. Tape or tie the ends of the string to the backs of the chairs.
3. Have the students blow up the balloon and twist the end so that it is easier to hold shut. Carefully tape the inflated balloon to the straw. Still holding the open end of the straw-balloon rocket closed, move it to the end of the string with the open end facing one of the chair backs.
4. Have the students let go of the open end of the straw-balloon rocket and watch what happens!

Explanation

The students should see the straw-balloon rocket takes off along the string towards the other chair. This is due to Newton's 3rd Law of Motion—for every action there is an equal, but opposite reaction. In other words, the force of the air leaving the balloon pushes the straw forward across the string. Rockets are launched with the same principles—the force of the fuel burning at the end of the rocket propels the rocket forward.

Take it further
Have the students repeat the demonstration, inflating the straw-balloon rocket to different levels to see how this affects the results.

Multi-Week Projects and Activities

Multi-week Projects
✂ **Solar System Model** - There is nothing to add to the solar system model this week.

Activities For This Week
✂ **Laws of Motion** – Have the students learn about the laws of motion by watching the following video:
🖱 https://www.youtube.com/watch?v=aA_mqSzbkM0

✂ **Rocket Launch** – Have the students build and launch a rocket. You will need a build-a-rocket kit, which you can purchase at Rainbow Resources or Amazon. We recommend getting one of the rocket kits by Estes that comes with the rocket pieces, a launch pad, and a rocket engine. Please follow the directions included in the kit you purchase.

Memorization

Copywork/Dictation
☞ **Copywork Sentence**
A spacewalk is a period of time spend outside of a spacecraft by an astronaut.

☞ **Dictation Selection**
A spacewalk is a period of time spend outside of a spacecraft by an astronaut. The first spacewalk was done by Alexi Leonov on March 18th, 1965. It only lasted 10 minutes. Today, astronauts do spacewalks to repair man-made satellites, to check on the outside of their spacecrafts, or to help build parts of the International Space Station.

Chapter 14 Notes

Chapter 15: Grid Schedule

	Supplies Needed
Demo	• Hard-boiled egg, Warm water, Bottle with large mouth (e.g., sports drink bottle), Access to a freezer
Projects	• Materials will vary based on the type of planet model you choose to make. • Plastic cup, Sharpie markers, Pan, Spray oil, Foil

Chapter Summary

The chapter opens with Blaine and Tracey playing the different versions of Planet Prowess after telling Arty about the new record on the game made by KIWI. After playing the VR version, the twins learn about dwarf planets. A few hours later, the games begin. The twins watch from the backstage as the top five gamers they met earlier are introduced, plus a new lucky local gamer, Kiwi Jones. The gamers begin on the 1981 version of the game: Wayne Hammer plays what he thinks will be a top score, until the leaderboard is revealed, and KIWI is in the number 1 spot and HAMM holds the number 3 spot. The gamers and the crowd are shocked at the revelation, as it was generally accepted that no one would beat the THIS score made by Robbie Thistler. The gamers continue to play, each with their own style. As the top gamers play, Arty Stone tells the twins about black holes. When it is Kiwi's turn, he sets another top record on the '81 version. Four gamers—Agnus the Librarian, *El Cohete Loco*, Wayne Hammer, and Kiwi Jones—move on to the 2001 version of the game. Kiwi sets another top record, but this time he doesn't beat Robbie Thistler's record. The chapter ends with only the top two gamers, Kiwi Jones and Wayne Hammer, moving on to play the new virtual-reality version.

Weekly Schedule

	Day 1	Day 2	Day 3	Day 4
Read	☐ Read the section entitled "Dwarf Planet Diversions" of Chapter 15 in *SSA Volume 6: Astronomy*.	☐ Read the section entitled "Black Hole Humor" of Chapter 15 in *SSA Volume 6: Astronomy*.	☐ (*Optional*) Read one or all of the assigned pages from the encyclopedia of your choice.	☐ (*Optional*) Read one of the additional books from your library. ☐ (*Optional*) Read about Pluto's demotion to dwarf planet.
Write	☐ Fill out a Astronomy Record Sheet on SL pg. 71 on dwarf planets. ☐ Go over the vocabulary word and enter it into the Astronomy Glossary on SL pg. 96.	☐ Fill out a Astronomy Record Sheet on SL pg. 72 on black holes. ☐ (*Optional*) Add observations to the Night Sky Journal Sheet on SL pg. 68.	☐ (*Optional*) Write narration on the Astronomy Notes Sheet on SL pg. 74. ☐ Add information learned from the demonstration on SL pg. 74.	☐ (*Optional*) Complete the copywork or dictation assignment and add it to the Astronomy Notes sheet on SL pg. 74. ☐ (*Optional*) Fill out the record sheet on SL pg. 76 for one of the projects. ☐ (*Optional*) Take Astronomy Quiz #7.
Do	☐ (*Optional*) Make the Pluto model.	☐ (*Optional*) Do the black hole art project.	☐ Do the demonstration entitled "Sucked In."	☐ Work on the Solar System Model.

Chapter 15: List Schedule

	Supplies Needed
Demo	• Hard-boiled egg, Warm water, Bottle with large mouth (e.g., sports drink bottle), Access to a freezer
Projects	• Materials will vary based on the type of planet model you choose to make. • Plastic cup, Sharpie markers, Pan, Spray oil, Foil

Chapter Summary

The chapter opens with Blaine and Tracey playing the different versions of Planet Prowess after telling Arty about the new record on the game made by KIWI. After playing the VR version, the twins learn about dwarf planets. A few hours later, the games begin. The twins watch from the backstage as the top five gamers they met earlier are introduced, plus a new lucky local gamer, Kiwi Jones. The gamers begin on the 1981 version of the game: Wayne Hammer plays what he thinks will be a top score, until the leaderboard is revealed, and KIWI is in the number 1 spot and HAMM holds the number 3 spot. The gamers and the crowd are shocked at the revelation, as it was generally accepted that no one would beat the THIS score made by Robbie Thistler. The gamers continue to play, each with their own style. As the top gamers play, Arty Stone tells the twins about black holes. When it is Kiwi's turn, he sets another top record on the '81 version. Four gamers—Agnus the Librarian, *El Cohete Loco*, Wayne Hammer, and Kiwi Jones—move on to the 2001 version of the game. Kiwi sets another top record, but this time he doesn't beat Robbie Thistler's record. The chapter ends with only the top two gamers, Kiwi Jones and Wayne Hammer, moving on to play the new virtual-reality version.

Essential To-Do's

Read
- ☐ Read the section entitled "Dwarf Planet Diversions" of Chapter 15 in *SSA Volume 6: Astronomy*.
- ☐ Read the section entitled "Black Hole Humor" of Chapter 15 in *SSA Volume 6: Astronomy*.

Write
- ☐ Fill out a Astronomy Record Sheet on SL pg. 71 on dwarf planets.
- ☐ Go over the vocabulary word and enter it into the Astronomy Glossary on SL pg. 96.
- ☐ Fill out a Astronomy Record Sheet on SL pg. 72 on black holes.
- ☐ Add information learned from the demonstration on SL pg. 74.

Do
- ☐ Do the demo entitled "Sucked In."
- ☐ Work on the Solar System Model.

Optional Extras

Read
- ☐ Read one or all of the assigned pages from the encyclopedia of your choice.
- ☐ Read one of the additional books from your library.
- ☐ Read about Pluto's demotion to dwarf planet.

Write
- ☐ Add observations to the Night Sky Journal Sheet on SL pg. 68.
- ☐ Write a narration on the Astronomy Notes Sheet on SL pg. 74.
- ☐ Complete the copywork or dictation assignment and add it to the Astronomy Notes sheet on SL pg. 74.
- ☐ Fill out the record sheet on SL pg. 76 for one of the projects.
- ☐ Take Astronomy Quiz #7.

Do
- ☐ Make the Pluto model.
- ☐ Do the black hole art project.

Chapter 15: The Planet Prowess Leaderboard

Science-Oriented Books

Living Book Spine
📖 Chapter 15 of *The Sassafras Science Adventures Volume 6: Astronomy*

Optional Encyclopedia Readings
- *Basher Science Astronomy* pg. 48 (Pluto), pg. 49 (Eris), pg. 85 (Black Hole)
- *Usborne Children's Encyclopedia* pg. 272 (Pluto and beyond), pp. 262-263 (The Sun)
- *DK First Space Encyclopedia* pp. 74-75 (Pluto), pp. 98-99 (Black holes)
- *Kingfisher Science Encyclopedia* pp. 410-411 (The Solar System's Minor Members), pg. 430 (section on Black Holes)

Additional Living Books
- 📖 *When Is a Planet Not a Planet?: The Story of Pluto* by Elaine Scott
- 📖 *Pluto: Dwarf Planet (Scholastic News Nonfiction Readers: Space Science)* by Christine Taylor-Butler
- 📖 *Dwarf Planets: Pluto, Charon, Ceres, and Eris (Amazing Science: Planets)* by Nancy Loewen and Jeff Yesh
- 📖 *Black Holes (A True Book: Space)* by Ker Than
- 📖 *A Black Hole Is Not a Hole* by Carolyn Cinami DeCristofano
- 📖 *What's Inside a Black Hole? Theories About Space Phenomena (Beyond the Theory: Science of the Future)* by Tom Jackson

Notebooking (SCIDAT Logbook Information)

This week, you can have the students complete a night sky journal sheet. You can also have them fill out the logbook sheets for the dwarf planets and black holes. Here is the information they could include:

Night Sky Journal
This week, you can have your students observe for any of the things you have studied in the past fifteen weeks for your night sky observations.

Astronomy Record Sheets
Dwarf Planets
Information Learned

⇨ A dwarf planet meets some, but not all, of the criteria to be considered a planet. These are celestial bodies that orbit the sun, have a nearly round shape, and are not satellites, meaning that they do not orbit around other planets.

⇨ As of 2019, we have five dwarf planets:

1. Pluto – This dwarf planet was found in 1930 and it is now considered a dwarf planet thanks to its small size. It is farther from the sun than any of the planets, which means it is super cold! In 2017, we finally got the very first close-up pictures of Pluto. And now, we know that it is a rocky planet with ice and red snow. It is smaller than Earth's moon. Pluto does have several moons, the largest of which is Charon. A day on Pluto lasts 153 hours and it takes about

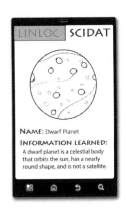

248 Earth years for Pluto to complete a lap around the sun.
2. *Eris – This dwarf planet is a bit larger than Pluto and is believed to be solid rock. It was discovered in 2003 and its appearance led to the famed IAU meeting where Pluto was demoted. Eris also has an elliptical orbit around the sun. A day on Eris lasts just under 26 hours, but it takes 557 Earth years to complete one lap around the sun!*
3. *Haumea – This dwarf planet is an astronomical oddball. It is shaped more like an oval-like football than a round sphere. Haumea orbits the sun just beyond Pluto. A day on Haumea only lasts about four hours, but it takes about 285 Earth years to make one lap around the sun.*
4. *Ceres – This dwarf planet is the only dwarf planet not found in the region known as the Kuiper belt, instead it is found between Mars and Jupiter. It was first spotted in 1801 and was thought to be an asteroid for many years until 2006, when the IAU said it met the new dwarf planet criteria. Ceres has a solid core that is covered with ice. A day on Ceres lasts around nine hours and it takes just over four and a half years for it to complete a lap around the sun.*
5. *Makemake – This dwarf planet was first spotted in 2005 and orbits the sun past where Pluto can be found. It is believed to have a surface similar to Pluto's. A day on Makemake lasts just under 23 hours and it takes about 305 years for it to complete a lap around the sun.*

BLACK HOLES
INFORMATION LEARNED

⇨ *Black holes are a bit of a mystery, but we do know that these are regions of space where the gravity is so strong that it sucks in everything, including light.*
⇨ *Black holes act like space whirlpools, not space vacuums.*
⇨ *Black holes are invisible, but we can spot them based on what goes on around them. Everything around them swirls around and gets sucked in. This creates a disc of particles that collects around the black hole.*
⇨ *Black holes also have a quasar, which is a jet of super-hot gas that shoots out above and below the black hole.*
⇨ *Scientist believe there are three different types of black holes:*
　1. *Stellar black holes – These are believed to form when a massive start dies.*
　2. *Super massive black holes – These are very large black holes that can be found at the center of a galaxy, including the Milky Way.*
　3. *Miniature black holes – This type of black hole has never actually been seen, but scientist do believe that smaller black holes can exist.*
⇨ *Albert Einstein initially predicted the existence of black holes with his theory of relativity, but it was Stephen Hawking who eventually used math to prove their existence.*

VOCABULARY

Have the older students look up the following term in the glossary in the Appendix on pp. 137-138 or in a science encyclopedia. Then, have them copy each definition onto a blank index card or into their SCIDAT logbook.

　✎ **DWARF PLANET** – A celestial body that looks like a planet but does not meet the three requirements to be one.

SCIENTIFIC DEMONSTRATION: SUCKED IN

MATERIALS
☑ Bottle with large mouth (**NOTE**—*The opening should be large enough for the egg to rest partially in the opening, but not so big that the egg falls into the bottle—a sports drink bottle works well for this.*)

- ☑ Hard-boiled egg, peeled
- ☑ Warm water
- ☑ Access to a freezer

Procedure
1. Fill the bottle with warm water and let it sit for several minutes.
2. Then, pour out the water and have the students set the egg in the mouth of the bottle. (**NOTE**—*Make sure that the wider part of the egg is the part that is placed in the opening.*)
3. Place the bottle into the freezer for about a minute. After the time has passed, have the students check the bottle out and observe what happens. (**NOTE**—*If the egg has not been sucked into the bottle at this point, leave the bottle in the freezer for several more minutes.*)

Explanation
The students should see the egg has been sucked down into the bottle. When the bottle was filled with water, it raised the temperature of the bottle. Even though the water was poured out, the air inside the bottle remained warm. When we placed the bottle in the freezer, the temperature inside the bottle decreased and since the egg was blocking the opening this caused the pressure inside the bottle to decrease. The drop in internal pressure created a vacuum that was able to pull the egg into the bottle. This vacuum action is not exactly the same as what happens in a black hole, but it is a fun way of picturing what happens out in space. Although scientists are not completely sure how black holes work, they do believe that they act more like whirlpools than vacuums. In other words, black holes spin around, pulling the matter near them, including light, into their centers.

Take it Further
Have the students get the egg back out! Flip the bottle over so that the egg is seated over the lid again. Then, hold the bottle under warm water until the egg pops out.

Multi-Week Projects and Activities

Multi-week Projects
- **Solar System Model** – This week, the students will add the dwarf planet, Pluto, to their solar system model. Have the students cut out the planet from copy paper or construction paper using the diameters below. Have them color the planet using a picture from the suggested books or from the Internet. When they are done, have them add the planet to their solar system using the distances below.
  - ⇨ Diameter (wall version): 1/2 in
  - ⇨ Distance (wall version): 13 ft, 3 in
  - ⇨ Diameter (lap version): 0.3 cm
  - ⇨ Distance (lap version): 79.6 cm

Activities For This Week
- **Pluto Model** – Have the students make a 3D model of Pluto. You can have them paint a Styrofoam ball, or you can have them make a paper-mâché model. (**NOTE**—*See the directions found on Appendix pg. 129 for the paper-mâché planet.*)
- **Pluto's Demotion** – If you are like me, you remember Pluto as the ninth planet. In 2006, Pluto was reclassified as a dwarf planet when the International Astronomical Union set out to define a planet. This is a good opportunity to discuss with the students how science is constantly changing as new discoveries are made and new information is found. Here are several links to articles on why Pluto was demoted:
  - This article explains the August 2006 decision and the opposition: http://www.msnbc.msn.com/id/14489259/152

- This article features Pluto, the planet that used to be: http://www.wired.com/science/discoveries/news/2001/01/41328
- This article explains why Pluto was demoted: http://www.space.com/scienceastronomy/060824_planet_definition.html

✂ **BLACK HOLE ART** – Have the students make an artistic representation of a black hole. You will need a plastic cup, Sharpie markers, a pan, spray oil, foil, and access to an oven. See the following post for directions:
- https://elementalscience.com/blogs/science-activities/how-to-create-a-black-hole-with-sharpies

MEMORIZATION

COPYWORK/DICTATION
☞ **COPYWORK SENTENCE**
Black holes act like space whirlpools, but we don't know much about them.

☞ **DICTATION SELECTION**
A dwarf planet meets some, but not all, of the criteria to be considered a planet. These are celestial bodies that orbit the sun, have a nearly round shape, and are not satellites, meaning that they do not orbit around other planets. As of 2019, we have five dwarf planets: Pluto, Eris, Haumea, Ceres, and Makemake.

QUIZ INFORMATION

This week, you can give the students a quiz based on what they learned in chapters 14 and 15. You can find this quiz in the Appendix on pg. 153.

QUIZ #7 ANSWERS
1. True
2. Hot, freezing
3. Escape velocity
4. Carry things into space, take people into space, burn fuel to get into space
5. False (*Pluto is very far from the sun.*)
6. Dwarf planets
7. Whirlpools
8. Albert Einstein, Stephen Hawking

CHAPTER 15 NOTES

Chapter 16: Grid Schedule

	Supplies Needed
Demo	• Foil, Toilet paper tube, Pin, Small flashlight, Constellation picture for Orion (Appendix pg. 134), Rubber band, Sharpie marker
Projects	• Marshmallows, Toothpicks • Gold star stickers, White crayon, Paper, Dark blue or black paint

Chapter Summary

The chapter opens with the end of the New Zealand Space Games. Wayne Hammer arrogantly defeats Kiwi Jones in the final virtual reality round. But then Robbie Thistler, who has been hiding in plain sight as Arty Stone, shows up and defeats the Hammer! The twins send in their SCIDAT info and zip off to the next location in Bollywood. They land on the set of the popular show, *Star Check*, in the make-up room. They overhear their local expert, Ravi, taking with the director, Varun, who is worried about *Star Check* being cancelled. The twins learn about constellations before they are mistaken for the Orion siblings. characters from the show who were kidnapped by an evil alien race called the Borbothians. Blaine and Tracey learn about the Orion constellation as they roll with the mistake and are quickly transformed into the Orion siblings. Tracey is disappointed by their drab garb, but they soon are out on the set meeting the cast. Ravi assures them all that if they put on the best performance, there is no way they will get cancelled. The chapter ends with the beginning of the first take of a scene from the next *Star Check* episode.

Weekly Schedule

	Day 1	Day 2	Day 3	Day 4
Read	☐ Read the section entitled "Constellation Display" of Chapter 16 in *SSA Volume 6: Astronomy*.	☐ Read the section entitled "Presenting Orion" of Chapter 16 in *SSA Volume 6: Astronomy*.	☐ (*Optional*) Read one or all of the assigned pages from the encyclopedia of your choice.	☐ (*Optional*) Read one of the additional books from your library.
Write	☐ Fill out a Astronomy Record Sheet on SL pg. 79 on constellations. ☐ Go over the vocabulary word and enter it into the Astronomy Glossary on SL pg. 97.	☐ Fill out a Astronomy Record Sheet on SL pg. 80 on Orion. ☐ (*Optional*) Add observations to the Night Sky Journal Sheet on SL pg. 77.	☐ (*Optional*) Write narration on the Astronomy Notes Sheet on SL pg. 83. ☐ Add information learned from the demonstration on SL pg. 83.	☐ (*Optional*) Complete the copywork or dictation assignment and add it to the Astronomy Notes sheet on SL pg. 83. ☐ (*Optional*) Fill out the record sheet on SL pg. 85 for one of the projects.
Do	☐ (*Optional*) Make a few constellation models.	☐ (*Optional*) Do the constellation resist project.	☐ Do the demonstration entitled "Flashlight Planetarium."	☐ Begin the constellation book.

Chapter 16: List schedule

	Supplies Needed
Demo	• Foil, Toilet paper tube, Pin, Small flashlight, Constellation picture for Orion (Appendix pg. 134), Rubber band, Sharpie marker
Projects	• Marshmallows, Toothpicks • Gold star stickers, White crayon, Paper, Dark blue or black paint

Chapter Summary

The chapter opens with the end of the New Zealand Space Games. Wayne Hammer arrogantly defeats Kiwi Jones in the final virtual reality round. But then Robbie Thistler, who has been hiding in plain sight as Arty Stone, shows up and defeats the Hammer! The twins send in their SCIDAT info and zip off to the next location in Bollywood. They land on the set of the popular show, *Star Check*, in the make-up room. They overhear their local expert, Ravi, taking with the director, Varun, who is worried about *Star Check* being cancelled. The twins learn about constellations before they are mistaken for the Orion siblings. characters from the show who were kidnapped by an evil alien race called the Borbothians. Blaine and Tracey learn about the Orion constellation as they roll with the mistake and are quickly transformed into the Orion siblings. Tracey is disappointed by their drab garb, but they soon are out on the set meeting the cast. Ravi assures them all that if they put on the best performance, there is no way they will get cancelled. The chapter ends with the beginning of the first take of a scene from the next *Star Check* episode.

Essential To-Do's

Read
☐ Read the section entitled "Constellation Display" of Chapter 16 in *SSA Volume 6: Astronomy*.
☐ Read the section entitled "Presenting Orion" of Chapter 16 in *SSA Volume 6: Astronomy*.

Write
☐ Fill out a Astronomy Record Sheet on SL pg. 79 on constellations.
☐ Go over the vocabulary word and enter it into the Astronomy Glossary on SL pg. 97.
☐ Fill out a Astronomy Record Sheet on SL pg. 80 on Orion.
☐ Add information learned from the demonstration on SL pg. 83.

Do
☐ Do the demonstration entitled "Flashlight Planetarium."
☐ Begin the constellation book.

Optional Extras

Read
☐ Read one or all of the assigned pages from the encyclopedia of your choice.
☐ Read one of the additional books from your library.

Write
☐ Add observations to the Night Sky Journal Sheet on SL pg. 77.
☐ Write a narration on the Astronomy Notes Sheet on SL pg. 83.
☐ Complete the copywork or dictation assignment and add it to the Astronomy Notes sheet on SL pg. 83.
☐ Fill out the record sheet on SL pg. 85 for one of the projects.

Do
☐ Make a few constellation models.
☐ Do the constellation resist project.

Chapter 16: The Set of *Star Check*

Science-Oriented Books

Living Book Spine
- Chapter 16 of *The Sassafras Science Adventures Volume 6: Astronomy*

Optional Encyclopedia Readings
- *Basher Science Astronomy* pg. 58 (Constellation)
- *Usborne Children's Encyclopedia* pg. 278-279 (Looking at the Night Sky)
- *DK First Space Encyclopedia* pp. 112-113 (Constellations)
- *Kingfisher Science Encyclopedia* pg. 396 (Constellations)

Additional Living Books
NOTE—These books can also be used for the next chapter.
- *Glow in the Dark Constellations* by C. E. Thompson
- *What We See in the Stars: An Illustrated Tour of the Night Sky* by Kelsey Oseid
- *Zoo in the Sky: A Book of Animal Constellations* by Jacqueline Mitton
- *Once Upon a Starry Night: A Book of Constellations* by Jacqueline Mitton and Christina Balit

Notebooking (SCIDAT Logbook Information)

This week, you can have the students complete a night sky journal sheet. You can also have them fill out the logbook sheets for constellations and the Orion constellation. Here is the information they could include:

Night Sky Journal
This week, you can have your students observe the constellations, especially Orion, for your night sky observations.

Astronomy Record Sheets
Constellations
Information Learned
- Constellations are a collection of stars that have been named. They are all visible from Earth.
- All the constellations have Latin names, and some have a local nickname, but all have a story behind their names.
- Many constellations were name after the characters found in ancient Greek myths.
- The stars within a constellation can 50 to 600 light years away from each other, even though they look like they are much closer to us on Earth.
- Astronomers once believed that these stars resided in a globe around the Earth, which they called the celestial sphere. Although we now know this is not true, it still is used as a helpful way of pinpointing the stars. The celestial sphere is cut into two halves – the northern celestial hemisphere and the southern celestial hemisphere. What constellations you see and when you see them depends upon where you live on Earth.

Orion
Information Learned

- *Orion is a well-known constellation because it is visibly to both the northern hemisphere, during winter, and southern hemisphere, during summer.*
- *This constellation is the shape of a man formed from about 20 stars, two of which – Rigel and Betelgeuse – are among the top ten brightest stars in the sky. The three stars in the center form Orion's belt, which is often still visible even in the city.*
- *The story behind the Orion constellation is rooted in Greek mythology.*
- *The story says that Orion was a great hunter who roamed the forest with his dog Sirius. One day he ran across seven beautiful sisters and, hoping to marry one of them, he ran after the girls. The girls were scared and fled to Zeus for help. He turned them all into birds and they flew off, leaving Orion all alone. The constellation is named after Orion, the lonely hunter, because it appears to move westward during the winter months along with its faithful Sirius, the Dog Star, close behind.*

Vocabulary
Have the older students look up the following term in the glossary in the Appendix on pp. 137-138 or in a science encyclopedia. Then, have them copy each definition onto a blank index card or into their SCIDAT logbook.
- **CONSTELLATION** – A group of stars that when viewed from Earth form the outline of an object or figure.

Scientific Demonstration: Flashlight Planetarium
Materials
- ☑ Foil
- ☑ Toilet paper tube
- ☑ Pin
- ☑ Small flashlight
- ☑ Constellation picture for Orion (Appendix pg. 134)
- ☑ Rubber band
- ☑ Sharpie marker

Procedure
(NOTE—*This flashlight planetarium will be used next week as well.*)
1. Have the students cut the foil into three squares so that each square will fit around the end of the toilet paper tube with enough leftover material to be secured by the rubber band. Set two of the sheets aside for use next week.
2. Have the students use a pin and the constellation picture on Appendix pg. 134 to create a foil constellations of Orion to use in their flashlight planetarium. Have them use a Sharpie maker to write the name of constellation in the bottom corner of the piece of foil.
3. Then, have the students view the constellation by placing the foil template at one end of the toilet paper tube so that the constellation is centered in the middle of the tube. Have them secure the foil in place using the rubber band.
4. Next, set the tube on a desk or counter and place the flashlight in the other end of the tube and aim the foil end of the tube towards a wall.
5. Finally, turn off the lights and turn on the flashlights to view the constellations!

EXPLANATION
The students should see outline of the constellation on the wall in front of them.

TAKE IT FURTHER
Have the students learn about the stories behind the Orion constellation they made for their flashlight planetarium. You can read about these online or in one of the following resources:
- *The Heavenly Zoo: Legends and Tales of the Stars* by Alison Lurie (NOTE—*This book is out of print, but it is wonderful if you can get your hands on it!*)
- *The Stars and Their Stories* by Alice Griffith (NOTE—*This book can be downloaded for free from Archive.org at the following link: https://archive.org/details/starsandtheirst00unkngoog*)

MULTI-WEEK PROJECTS AND ACTIVITIES

MULTI-WEEK PROJECTS
- **CONSTELLATION BOOK** – Over the next two weeks, have the students make a book of the constellations they study. This week, have them make a page about what constellations are and a page for the Orion constellation. They can use the constellation resist they make this week for the cover of the book.

ACTIVITIES FOR THIS WEEK
- **CONSTELLATION MODELS** – Have the students make a model of several different constellations, either the ones in the novel (or logbook) or ones from the additional encyclopedias. You will need marshmallows and toothpicks for this project.
- **CONSTELLATION RESIST** – Have the students make a constellation model using resist painting. You will need gold star stickers, white crayon, paper, and dark blue paint. Have the students make the outline of the Orion constellation, using the star stickers and white crayon. After they have done this, have them paint all over it with the dark blue paint. Once the paint is dry, have the students remove the star stickers to reveal their full constellations.

MEMORIZATION

COPYWORK/DICTATION
- **COPYWORK SENTENCE**
  Constellations are a collection of stars that have been named. There are 88 constellations.
- **DICTATION SELECTION**
  Constellations are a collection of stars that have been named. There are 88 internationally recognized constellations, all of which are visible from Earth. All the constellations have a Latin name and some have a local nickname. Many constellations were name after the characters found in ancient Greek myths.

Chapter 16 Notes

Chapter 17: Grid Schedule

	Supplies Needed
Demo	• Flashlight planetarium from previous week
Projects	• Materials will vary based on what you decided to do for your constellation party.

Chapter Summary

The chapter opens with the arrival of the two executives who are watching the production of *Star Check* to determine if the show is worth keeping. Ravi introduces himself, explains the background for the scene they are working on, and assures them that *Star Check* is worth their time and money. Next, the crew gets down to business. They do take after take of the scene, some slower, some faster, some placing emphasis on different emotions. Blaine and Tracey hold their own through the takes and manage to learn about Ursa Major and Ursa Minor at the same time. The cast finally gets the scene just right and the executives let them know that *Star Check* will continue. The chapter ends with what all things in Bollywood end with – a dance party!

Weekly Schedule

	Day 1	Day 2	Day 3	Day 4
Read	☐ Read the section entitled "Ursa Major Fanfare" of Chapter 17 in *SSA Volume 6: Astronomy*.	☐ Read the section entitled "Viewing Ursa Minor" of Chapter 17 in *SSA Volume 6: Astronomy*.	☐ *(Optional)* Read one or all of the assigned pages from the encyclopedia of your choice.	☐ *(Optional)* Read one of the additional books from your library.
Write	☐ Fill out a Astronomy Record Sheet on SL pg. 81 on Ursa Major.	☐ Fill out a Astronomy Record Sheet on SL pg. 82 on Ursa Minor. ☐ *(Optional)* Add observations to the Night Sky Journal Sheet on SL pg. 78.	☐ *(Optional)* Write narration on the Astronomy Notes Sheet on SL pg. 84. ☐ Add information learned from the demonstration on SL pg. 84.	☐ *(Optional)* Complete the copywork or dictation assignment and add it to the Astronomy Notes sheet on SL pg. 84. ☐ *(Optional)* Fill out the record sheet on SL pg. 86 for one of the projects. ☐ *(Optional)* Take Astronomy Quiz #8.
Do	☐ Make a page for Ursa Major for their constellation book.	☐ Make a page for Ursa Minor for their constellation book.	☐ Do the demonstration entitled "Flashlight Planetarium."	☐ *(Optional)* Have a constellation party.

Chapter 17: List Schedule

	Supplies Needed
Demo	• Flashlight planetarium from previous week
Projects	• Materials will vary based on what you decided to do for your constellation party.

Chapter Summary

The chapter opens with the arrival of the two executives who are watching the production of *Star Check* to determine if the show is worth keeping. Ravi introduces himself, explains the background for the scene they are working on, and assures them that *Star Check* is worth their time and money. Next, the crew gets down to business. They do take after take of the scene, some slower, some faster, some placing emphasis on different emotions. Blaine and Tracey hold their own through the takes and manage to learn about Ursa Major and Ursa Minor at the same time. The cast finally gets the scene just right and the executives let them know that *Star Check* will continue. The chapter ends with what all things in Bollywood end with – a dance party!

Essential To-Do's

Read
☐ Read the section entitled "Ursa Major Fanfare" of Chapter 17 in *SSA Volume 6: Astronomy*.
☐ Read the section entitled "Viewing Ursa Minor" of Chapter 17 in *SSA Volume 6: Astronomy*.

Write
☐ Fill out a Astronomy Record Sheet on SL pg. 81 on Ursa Major.
☐ Fill out a Astronomy Record Sheet on SL pg. 82 on Ursa Minor.
☐ Add information learned from the demonstration on SL pg. 84.

Do
☐ Do the demo entitled "Flashlight Planetarium."
☐ Finish the constellation book with pages for Ursa Major and Ursa Minor.

Optional Extras

Read
☐ Read one or all of the assigned pages from the encyclopedia of your choice.
☐ Read one of the additional books from your library.

Write
☐ Add observations to the Night Sky Journal Sheet on SL pg. 78.
☐ Write a narration on the Astronomy Notes Sheet on SL pg. 84.
☐ Complete the copywork or dictation assignment and add it to the Astronomy Notes sheet on SL pg. 84.
☐ Fill out the record sheet on SL pg. 86 for one of the projects.
☐ Take Astronomy Quiz #8.

Do
☐ Have a constellation party.

Chapter 17: The Orion Siblings in Bollywood

Science-Oriented Books

Living Book Spine
- Chapter 17 of *The Sassafras Science Adventures Volume 6: Astronomy*

Optional Encyclopedia Readings
- *Basher Science Astronomy* (No pages scheduled)
- *Usborne Children's Encyclopedia* (No pages scheduled)
- *DK First Space Encyclopedia* pp. 114-115 (The Northern Sky), pp. 116-117 (The Southern Sky)
- *Kingfisher Science Encyclopedia* pg. 397 (Looking at the Constellations)

Additional Living Books
NOTE—These books were also scheduled for use in the previous chapter.
- *Glow in the Dark Constellations* by C. E. Thompson
- *What We See in the Stars: An Illustrated Tour of the Night Sky* by Kelsey Oseid
- *Zoo in the Sky: A Book of Animal Constellations* by Jacqueline Mitton
- *Once Upon a Starry Night: A Book of Constellations* by Jacqueline Mitton and Christina Balit

Notebooking (SCIDAT Logbook Information)

This week, you can have the students complete a night sky journal sheet. You can also have them fill out the logbook sheets for the Ursa Major and Ursa Minor constellations. Here is the information they could include:

Night Sky Journal
This week, you can have your students observe the constellations, especially Ursa Major and Ursa Minor, for your night sky observations.

Astronomy Record Sheets
Ursa Major
Information Learned
- Ursa Major is also known as the Great Bear, the Big Dipper, or the Plough.
- It is made up of seven stars: four in its bowl and three in its handle.
- This constellation is very helpful for locating other constellations. In the northern hemisphere, it is visible year-round, so it is often used for navigating and orienting the night sky.
- The story behind the Ursa Major constellation is rooted in Greek mythology.
- The story says that Hera, the queen of the gods, was jealous of a beautiful young woman named Callisto. She made a plan to hurt Callisto, but Zeus, the king of the gods, found out about the plan. To keep Callisto safe, he changed her into a bear and eventually set her into the sky.

Ursa Minor
Information Learned
- Ursa Minor is also known as the Little Bear or Little Dipper.
- It is also made up of seven stars: four in its bowl and three in its handle, one of which is Polaris.

⇨ *Polaris, also known as the North Star or the Pole Star, is the brightest star in the Little Dipper constellation, but not the brightest in the sky. Polaris can be found at the tip of the Little Dipper's handle and if you follow it in a line to the left, you will find the edge of the bowl of the Big Dipper.*

⇨ *The story behind the Ursa Minor constellation is also rooted in Greek mythology.*

⇨ *The story says that one day Callisto's son, Arcas, was out hunting when he came across a bear, who was really his mother. Zeus saw the encounter, and before Arcas could shoot he turned the boy into a small bear and eventually placed him in the sky along with his mother. Since a little bear is always found with its mother, Ursa Minor is said to always be near Ursa Major.*

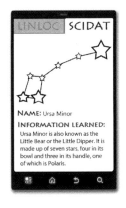

Vocabulary
There is no vocabulary to add this week.

Scientific Demonstration: Flashlight Planetarium

Materials
☑ Flashlight planetarium from previous week

Procedure
1. Have the students get the two of foil sheets they aside last week. Then have them use a pin and the constellation picture on Appendix pg. 134 to create foil constellations of Ursa Major and Ursa Minor to use in their flashlight planetarium. Have them use a Sharpie maker to write the name of the constellations in the bottom corner of the pieces of foil.
2. Then, have the students view each of the constellation, by placing the foil template at one end of the toilet paper tube so that the constellation is centered in the middle of the tube. Have them secure the foil in place using the rubber band.
3. Next, set the tube on a desk or counter and place the flashlight in the other end of the tube and aim the foil end of the tube towards a wall.
4. Finally, turn off the lights and turn on the flashlights to view the constellations!

Explanation
The students should see outline of the constellation on the wall in front of them.

Take it further
Have the students learn about the stories behind the Ursa Major and Ursa Minor constellation they made for their flashlight planetarium. You can read about these online or in one of the following resources:

📖 *The Heavenly Zoo: Legends and Tales of the Stars* by Alison Lurie (**NOTE**—*This book is out of print, but it is wonderful if you can get your hands on it!*)

📖 *The Stars and Their Stories* by Alice Griffith (**NOTE**—*This book can be downloaded for free from Archive.org at the following link: https://archive.org/details/starsandtheirst00unkngoog*)

Multi-Week Projects and Activities

Multi-week Projects
✂ **Constellation Book** – This week, the students will finish their constellation book about the constellations they study. Have them make pages for Ursa Major and Ursa Minor constellations.

Activities For This Week
✂ **Constellation Party** – Have a constellation party with the students. You can make

constellation cupcakes from the recipe below:
  http://www.piecesbypolly.com/2012/02/easy-constellation-cupcakes-space-party.html

You can also play the "Which constellation is it?" game. For this, use the constellation cards on pg. 134 of the Appendix and cut the name off the card. Have the students lay out the constellation pictures and then try to match the names with the pictures.

Memorization

Copywork/Dictation
☞ **Copywork Sentence**
Ursa Major is also called the Big Dipper. Ursa Minor is also called the Little Dipper.

☞ **Dictation Selection**
Ursa Major is also known as the Great Bear, the Big Dipper, or the Plough. It is made up of seven stars: four in its bowl and three in its handle. Ursa Minor is also known as the Little Bear or Little Dipper. It is also made up of seven stars: four in its bowl and three in its handle, one of which is Polaris.

Quiz Information

This week, you can give the students a quiz based on what they learned in chapters 16 and 17. You can find this quiz in the Appendix on pg. 155.

Quiz #8 Answers
1. Constellation
2. Latin
3. True
4. Students can answer - Great Bear, Plough, or Big Dipper
5. Students can answer - Polaris or the North Star
6. B, A, C

Chapter 17 Notes

Chapter 18: Grid Schedule

	Supplies Needed
Demo	• Planetary Bingo Cards (Download for free from Elemental Science)
Projects	• Materials will vary based on what you decided to do for the alien craft.

Chapter Summary

The chapter opens with Uncle Cecil putting Socrates and Aristotle in charge while he runs down to the Left-Handed Turtle Market for some foil. On his way, he spots some very strange unidentified flying objects. We switch to Blaine and Tracey who are dancing their hearts out in Bollywood. We then switch to Summer, Ulysses, and President Lincoln who are taking the heliquickter back to Cecil's after wrapping up their business with Paul Sims at the National Air and Space Museum. They find Cecil's sticky notes and run off to meet him at the market. We then switch to the Triple S agents, who are currently descending upon the Left-Handed Turtle Market to capture Adrianna Archer. Back to Blaine and Tracey who are opening up the LINLOC app and zipping back to Uncle Cecil's to wrap up their astronomy leg and get their bonus data. They arrive to find the basement empty. The bonus data arrives just as they find Cecil's notes, so the twins also head off to the market. Back at the Left-Handed Turtle, Preston is trying to deal with a very persistent woman (it's really Adrianna Archer) who insists on setting up a make-up stand by the cash register. Moments later, it's as if an invisible person walks into the market and now we know that the Man With No Eyebrows is there, too. Outside, the Triple S is verifying the target and coming up with a plan to apprehend Adrianna. Back inside, Adrianna has gotten Cecil into her "makeup station" chair and is about to "transform" him when Summer walks up. Summer reveals who the makeup artist is, just as the Triple S Agent, Evan DeBlose, walks in to greet Adrianna. Blaine and Tracey walk in just as the Triple S is taking her out – they miss all the action. We find out that the Man With No Eyebrows has silently watched the whole thing and now he is ready to refocus his plans. The chapter ends with a twist – we see Paul Sims paying a bond that releases the Rotary Club from back in Chapter 11!

Weekly Schedule

	Day 1	Day 2	Day 3	Day 4
Read	☐ Read the section entitled "Bonus Data" of Chapter 18 in *SSA SSA Volume 6: Astronomy*.	☐ (*Optional*) Read one or all of the assigned pages from the encyclopedia of your choice.	☐ Read the section entitled "Invasion of the Pecan Street Neighborhood" of Chapter 18 in *SSA Volume 6: Astronomy*.	☐ Read one of the additional books from your library.
Write	☐ Add information learned about UFOs to the SL pg. 87.	☐ (*Optional*) Write a narration on the Astronomy Notes Sheet on SL pg. 87. ☐ Add information learned from the demonstration on SL pg. 88.	☐ (*Optional*) Complete the copywork or dictation assignment and add it to the Astronomy Notes sheet on SL pg. 88.	
Do	☐ (*Optional*) Make an alien craft.	☐ Play a game of Planetary Bingo.		☐ Review the work you have done over the unit.

Chapter 18: List Schedule

Supplies Needed	
Demo	• Planetary Bingo Cards (Download for free from Elemental Science)
Projects	• Materials will vary based on what you decided to do for the alien craft.

Chapter Summary

The chapter opens with Uncle Cecil putting Socrates and Aristotle in charge while he runs down to the Left-Handed Turtle Market for some foil. On his way, he spots some very strange unidentified flying objects. We switch to Blaine and Tracey who are dancing their hearts out in Bollywood. We then switch to Summer, Ulysses, and President Lincoln who are taking the heliquickter back to Cecil's after wrapping up their business with Paul Sims at the National Air and Space Museum. They find Cecil's sticky notes and run off to meet him at the market. We then switch to the Triple S agents, who are currently descending upon the Left-Handed Turtle Market to capture Adrianna Archer. Back to Blaine and Tracey who are opening up the LINLOC app and zipping back to Uncle Cecil's to wrap up their astronomy leg and get their bonus data. They arrive to find the basement empty. The bonus data arrives just as they find Cecil's notes, so the twins also head off to the market. Back at the Left-Handed Turtle, Preston is trying to deal with a very persistent woman (it's really Adrianna Archer) who insists on setting up a make-up stand by the cash register. Moments later, it's as if an invisible person walks into the market and now we know that the Man With No Eyebrows is there, too. Outside, the Triple S is verifying the target and coming up with a plan to apprehend Adrianna. Back inside, Adrianna has gotten Cecil into her "makeup station" chair and is about to "transform" him when Summer walks up. Summer reveals who the makeup artist is, just as the Triple S Agent, Evan DeBlose, walks in to greet Adrianna. Blaine and Tracey walk in just as the Triple S is taking her out – they miss all the action. We find out that the Man With No Eyebrows has silently watched the whole thing and now he is ready to refocus his plans. The chapter ends with a twist – we see Paul Sims paying a bond that releases the Rotary Club from back in Chapter 11!

Essential To-Do's

Read
☐ Read the section entitled "Bonus Data" of Chapter 18 in SSA* *Volume 6: Astronomy*.
☐ Read the section entitled "Invasion of the Pecan Street Neighborhood" of Chapter 18 in SSA* *Volume 6: Astronomy*.

Write
☐ Add information learned about UFOs to the SL pg. 87.
☐ Add information learned from the demonstration on SL pg. 88.

Do
☐ Play a game of Planetary Bingo.
☐ Review the work you have done over the unit.

Optional Extras

Read
☐ Read one or all of the assigned pages from the encyclopedia of your choice.
☐ Read one of the additional books from your library.

Write
☐ Write a narration on the Astronomy Notes Sheet on SL pg. 87.
☐ Complete the copywork or dictation assignment and add it to the Astronomy Notes sheet on SL pg. 88.

Do
☐ Make an alien craft

Chapter 18: Back to Uncle Cecil's

Science-Oriented Books

Living Book Spine
- Chapter 18 of *The Sassafras Science Adventures Volume 6: Astronomy*

Optional Encyclopedia Readings
- *Basher Science Astronomy* (No pages scheduled)
- *Usborne Children's Encyclopedia* pp. 256-257 (Is Anyone Out There?)
- *DK First Space Encyclopedia* pp. 90-91 (UFOs)
- *Kingfisher Science Encyclopedia* (No pages scheduled)

Additional Living Books
NOTE—These books are just for fun!
- *The Three Little Aliens and the Big Bad Robot* by Margaret McNamara and Mark Fearing
- *Aliens in Underpants Save the World* (The Underpants Books) by Claire Freedman and Ben Cort
- *Your Alien* by Tammi Sauer and Goro Fujita

Notebooking (SCIDAT Logbook Information)

This week, you can have the students fill the Astronomy Notes sheets with the bonus data. Here's the information they could include:

Bonus Data
UFOs
- UFO stands for Unidentified Flying Objects—these are things that we see in the sky that we can't readily explain.
- Some of these sightings are weather balloons or new top-secret technology. Some can't be as easily explained.
- In 1947, people found debris that they claimed were from a flying saucer, or UFO, in Roswell, New Mexico. The Air Force said that it was a top-secret weather balloon.
- Today Roswell is home to the International UFO Museum and Research Center.

Vocabulary
There is no vocabulary to add this week.

Scientific Demonstration: Planetary Bingo

Materials
☑ Planetary Bingo from Elemental Science

Procedure
1. Download the game templates from the following website:
   - https://elementalscience.com/collections/free-printable-games/products/astronomy-game-planetary-bingo-free-ebook
2. Play Planetary Bingo according to the directions included in the game packet.

Multi-Week Projects and Activities

Multi-week Projects
- **Solar System Model** – Just for fun, you can have the students add a UFO to their solar system model. If you do this, let them choose the design and location placement for the UFO.

Activities For This Week
- **Alien Ships** – Have the students design and build their own alien craft. You can let their imaginations run wild for this project! Here are a few ideas to get those creative juices flowing:
  - https://www.momsandmunchkins.ca/alien-crafts/

Memorization

Copywork/Dictation
- **Copywork Sentence**

  UFO stands for Unidentified Flying Objects.

- **Dictation Selection**

  UFO stands for Unidentified Flying Objects. These are things that we see in the sky that we can't readily explain. Some of these sightings are weather balloons or new top-secret technology. Some can't be as easily explained.

Chapter 18 Notes

APPENDIX

LAB REPORT SHEET

Title
_____

Hypothesis (What I Think Will Happen)
_____
_____
_____

Materials (What We Used)

_____   _____
_____   _____
_____   _____
_____   _____

Procedure (What We Did)
_____
_____
_____
_____
_____
_____
_____
_____
_____
_____
_____

Observations and Results (What I Saw and Measured)
_____
_____
_____

Conclusion (What I Learned)

Paper Mâché Planet Model Directions

Supplies
You will need:
- ☑ Balloon
- ☑ Newspaper
- ☑ 1 Cup of flour
- ☑ ½ Cup of water
- ☑ 2 Tablespoons of salt
- ☑ Globe

Directions
1. Begin by having the students blow up the balloon.
2. Next, have them tear the newspaper into strips. As they are working on the newspaper strips, use the flour, water, and salt to make a thick paste. You can add more or less water to gain the desired consistency.
3. Then, have the students dip the strips into the paste mixture and cover the balloon with one layer.
4. Wait 30 minutes before having them add a second layer. As they do this, have them look at the globe to add any topographical features (e.g., mountains) to their model Earth.
5. Finally, set the paper mâché models in a warm, moisture-less location to dry out in preparation for next week.

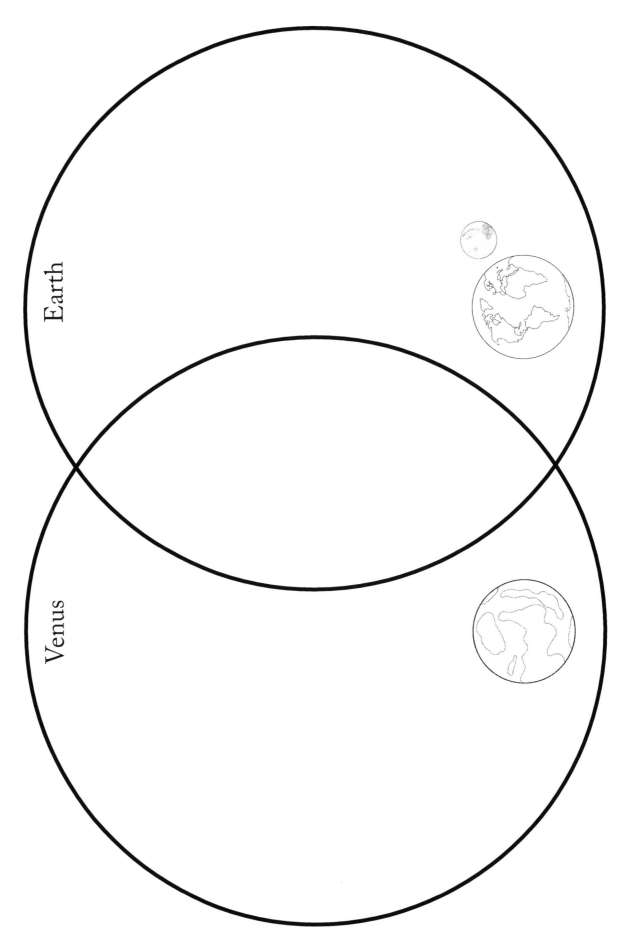

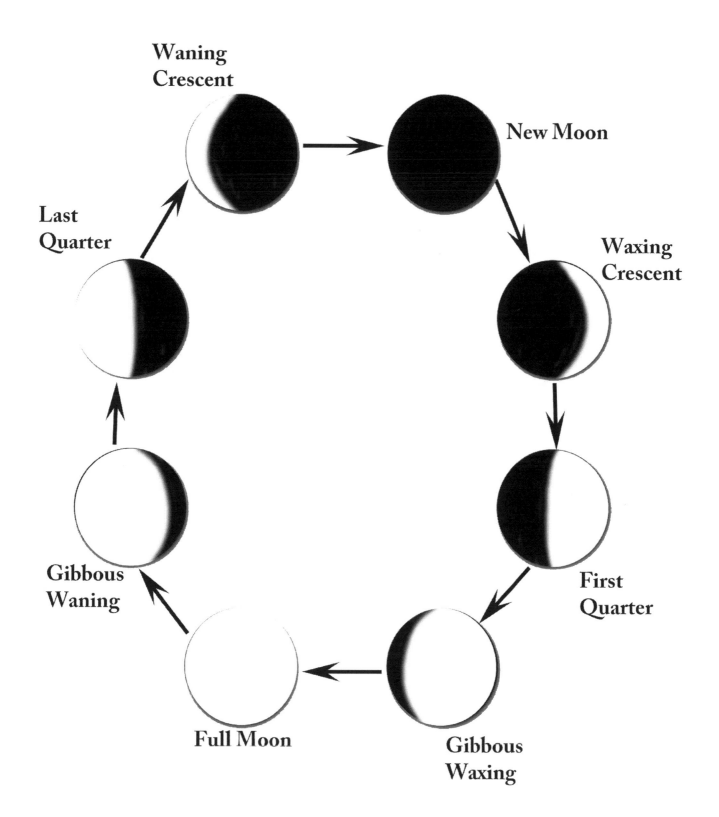

Scientist Biography Questionnaire

Which scientist did you choose?
_____

Title of the book you read
_____

When and where was the scientist born?
_____
_____
_____
_____

What was the scientist's major scientific contribution?
_____
_____
_____
_____
_____
_____

List the events that surround the scientist's major discovery.
_____
_____
_____
_____
_____
_____
_____
_____

List some other interesting events in the scientist's life.

_____
_____
_____
_____
_____
_____
_____
_____
_____

Why do you think that it is important to learn about this scientist?

_____
_____
_____
_____
_____

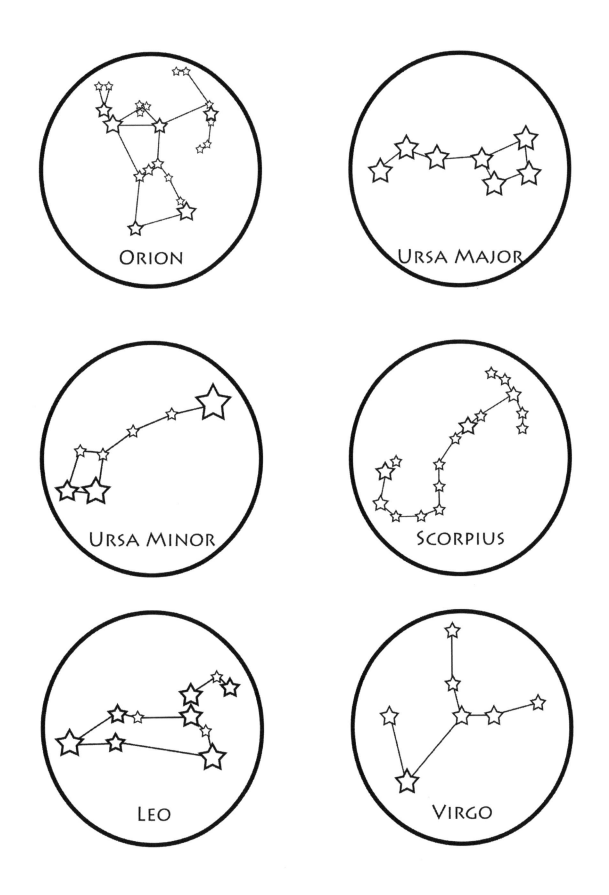

GLOSSARY

Astronomy Glossary

A

- ASTEROID – A rock orbiting the sun.
- ASTRONAUT – A person who travels into space.
- ASTRONOMER – A person who studies space and the things found in it.
- ASTRONOMY – The branch of science that studies what is out in space.
- ATMOSPHERE – A layer of gas that surrounds a planet.

B

C

- CONSTELLATION – A group of stars that when viewed from Earth form the outline of an object or figure.

D

E

- ESCAPE VELOCITY – The speed an object must reach in order to escape the force of Earth's gravity.

F

- FORCE – A push or pull on an object.

G

- GAS GIANT – A large planet in our solar system that is composed mainly of gas.
- GRAVITY – The force that pulls an object towards another larger object.

H

I

J

K

L

M

- METEOR – A rock that travels through space and burns up when it enters a planet's atmosphere; also known as a shooting star.
- MOON – A natural satellite in orbit around a planet.

N

O

- ORBIT – The path of an object in space.

P

- PLANET – A large ball of rock or gas that travels around a star.

Q

R

- REFLECTION – The change in direction of light rays that occurs when it hits an object and bounces off.
- REFRACTION – The bending of light as it passes through a different substance.

S

- SATELLITE – An object, either man-made or natural, that orbits something bigger than itself.
- SOLAR ECLIPSE – The time when the moon blocks the sun.
- SOLAR SYSTEM – A group of planets and other objects all in orbit around the sun.
- SOLAR WIND – A stream of tiny particles that blow off the sun and into space.
- SPACE STATION – A man-made structure that is launched into space and orbits around the sun-orbiting earth.
- STAR – A huge ball of exploding gas out in space.

T

- TELESCOPE – An instrument to look at things far away in space.

U

- UNIVERSE – The collection of all the matter, space, and energy that exists.

V

W

X

Y

Z

QUIZZES

Astronomy Quiz #1
Chapters 2 and 3

1. Our solar system includes:

   The sun   The 8 planets   The moons

         Asteroids   Comets

2. _____ from the sun keeps all these objects in our solar system orbiting around it.

3. Our solar system includes two bands of drifting _____ called the Asteroid Belt and the Kuiper Belt.

4. An asteroid is a _____ orbiting the sun.

   planet

   rock

   star

5. A star is really a huge ball of exploding _____.

               gas   water   air

6. Put the life cycle of a star in order from the birth of a start to the end of its life.

    ____ Explodes and shines.

    ____ Born in a nebula.

    ____ Becomes a white dwarf.

    ____ Grows hotter and hotter.

    ____ Burns out and begins to die.

7. **True or False:** As of right now, the International Space Station is the most expensive thing man has ever built.

8. The International Space Station, also known as the I.S.S., is made from several _____ that clip together.

Astronomy Quiz #2
Chapters 4 and 5

1. What are the four inner planets?

    M _____

    V _____

    E _____

    M _____

2. Mercury is _____ size of Earth.

        3 times the   one-third of the   the same

3. Venus has a thick atmosphere made up of

    _____.

    oxygen

    nitrogen

    carbon dioxide

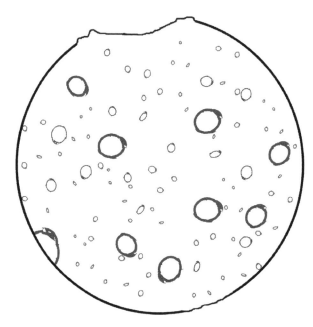

4. **True or False:** Earth is the closest planet to the sun.

5. Earth is the only planet that we know of that can support _____.

6. The surface of _____ looks like a great big desert of red, iron-rich dust with volcanoes and canyons dotting the landscape.

7. **True or False:** An orbit is the path that an object takes in space.

Astronomy Quiz #3
Chapters 6 and 7

1. What are the four outer planets?

    J _____

    S _____

    U _____

    N _____

2. Jupiter is the _____ planet in the solar system.

    smallest   largest

3. **True or False:** Jupiter's red spot is not moving in any way.

4. Saturn's rings are made up of _____.

   rocks, dust, and ice   gas, methane, and water   fire, dirt, and air

5. Saturn is a very _____ planet.

   heavy   average   light

6. True or False: Uranus has no moons.

7. True or False: Both Uranus and Neptune are known as gas giants.

8. _____ has the worst storms of the solar system.

Astronomy Quiz #4
Chapters 8 and 9

1. Match the type of galaxy with its description.

    ____ Spiral galaxies

    ____ Elliptical galaxies

    ____ Barred spiral galaxies

    ____ Irregular galaxies

    A. Round or oval shape with a bulge in the center, but no disk

    B. Hodgepodge of shapes, basically anything that is not spiral or elliptical, meaning they have no regular shape

    C. Pinwheel shape with a bulge and thin disk in the center

    D. Pinwheel shape with a bar of gas, dust, and stars running through the center

2. Our solar system is located in the _____ galaxy.

    C590   Milky Way

3. Telescopes make everything look

    _____.

    smaller   the same

    bigger

The Sassafras Guide to Astronomy ~ Quizzes

4. When light is refracted, it _____.

5. **True or False:** Satellites visit other planets, while space probes orbit the earth.

6. What can satellites do?

        Track weather   Give exact position

        Help communicate   Learn about space

7. A space probe is _____ spacecraft that collects information about objects in space.

8. When light is reflected, it hits a surface and _____ off.

Astronomy Quiz #5
Chapters 10 and 11

1. The sun is a _____.

2. **True or False:** Solar wind is a bunch of tiny particles that blow off of the sun.

3. During the day, the Earth faces (away from / towards) the sun.

   During the night, the Earth faces (away from / towards) the sun.

4. **True or False:** A full day-night cycle takes about 12 hours.

5. A _____ eclipse is when the moon blocks the sun.

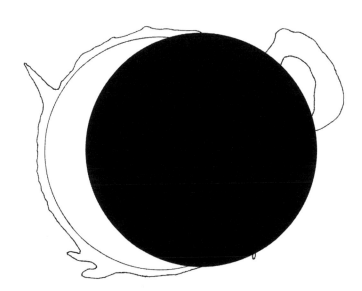

6. A _____ eclipse is when the Earth is between the sun and the moon.

7. **True or False:** Our closest neighbor in space is the moon.

8. The (full moon / new moon) is when the moon seems to disappear from the night sky. The (full moon / new moon) is when the whole moon appears to be lit.

Astronomy Quiz #6
Chapters 12 and 13

1. An astronaut is someone who travels _____.

   around the beltway   into space   under water

2. Match the parts of the space shuttle with what it does.

   ____ Boosters

   ____ Fuel tank

   ____ Shuttle

   A. Contains the bulk of the fuel for the main engine

   B. Provides the first bit of fuel needed before falling away back to earth

   C. Holds the compartments for the crew

3. Astronauts train _____ because it's the nearest thing to being weightless on Earth.

   underground

   underwater

4. **True or False:** The space shuttle is a reusable aircraft that can travel to space.

The Sassafras Guide to Astronomy ~ Quizzes   151

5. **True or False:** Comets are large balls of cotton and sand.

6. A comet's tail always points _____ the sun.

7. Match the ancient astronomer with what he is famous for.

____ Claudius Ptolemy

____ Nicholas Copernicus

____ Johannes Kepler

____ Galileo Galilei

A. He proposed that the sun, not the earth, was at the center of our universe.

B. He wrote several scientific books, including one that detailed his work on cataloguing the stars and the positions of the planets.

C. He discovered Jupiter's moons and invented the first telescope that was capable of magnifying things by 20 times.

D. He is best known for his three laws that explain the relationship between a planet's distance from the sun and the length of its orbit.

Astronomy Quiz #7
Chapters 14 and 15

1. **True or False:** A space walk is a period of time spend outside of a spacecraft by an astronaut.

2. When the space walk faces the sun, the astronaut gets _____, but as soon as they travel into the earth's shadow, the astronaut is _____.

3. To escape earth's gravity, the rocket has to ready over 7 miles per second, which is known as the _____ _____.

4. Pluto, Eris, and Ceres are all _____.

    dwarf planets

    planets

    stars

5. Rockets _____.

    carry things into space   take people into space

    burn fuel to get into space   don't go anywhere

6. **True or False:** Pluto is very close to the Sun.

7. Scientists believe that black holes act like space _____.

8. (Albert Einstein / Stephen Hawking) initially predicted the existence of black holes with his theory of relativity, but it was (Albert Einstein / Stephen Hawking) who eventually used math to prove their existence.

Astronomy Quiz #8
Chapters 16 and 17

1. A _____ is a pattern in the stars.

2. All constellations have a _____ name.

3. **True or False:** Orion can be seen from the northern and southern hemisphere at different times of the year.

4. Ursa major is also known as the _____.

5. Ursa minor contains _____ in its handle.

6. Match the constellation picture with its name.

  ____ Ursa Major

  A.
  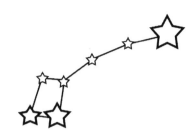

  ____ Ursa Minor

  B.

  C.

  ____ Orion
  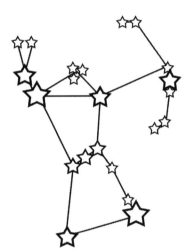

Made in the USA
Middletown, DE
10 February 2022